Inspiring | Educating | Creating | Entertaining

Brimming with creative inspiration, how-to projects, and useful information to enrich your everyday life, Quarto Knows is a favorite destination for those pursuing their interests and passions. Visit our site and dig deeper with our books into your area of interest: Quarto Creates, Quarto Cooks, Quarto Homes, Quarto Lives, Quarto Drives, Quarto Explores, Quarto Gifts, or Quarto Kids.

© 2019 Quarto Publishing Group USA Inc.
Text © 2019 Lisa Eldred Steinkopf

First published in 2019 by Cool Springs Press,
an imprint of The Quarto Group,
100 Cummings Center Suite 265D,
Beverly, MA 01915 USA.
T (978) 282-9590
F (978) 283-2742
www.QuartoKnows.com

Cool Springs Press titles are also available at discount for retail, wholesale, promotional, and bulk purchase. For details, contact the Special Sales Manager by email at specialsales@quarto.com or by mail at The Quarto Group, Attn: Special Sales Manager, 100 Cummings Center Suite 265D, Beverly, MA 01915 USA.

10 9 8 7 6 5 4 3 2 1

ISBN: 978-0-7603-6451-2

Digital edition published in 2019
eISBN: 978-0-7603-6452-9

Library of Congress Cataloging-in-Publication Data

Names: Steinkopf, Lisa Eldred, 1966- author.
Title: Grow in the dark : how to choose and care for low-light
    houseplants / Lisa Eldred Steinkopf.
Description: Beverly, MA : Cool Springs Press, 2019. | Includes index.
Identifiers: LCCN 2018050845 | ISBN 9780760364512 (paper over board)
Subjects: LCSH: House plants. | Plants--Effect of light on.
Classification: LCC SB419 .S7274 2019 | DDC 635.9/65--dc23
LC record available at https://lccn.loc.gov/2018050845

Acquiring Editor: Alyssa Bluhm
Art Director: Cindy Samargia Laun
Photography: Heather Saunders
Cover and Page Design: Evelin Kasikov

Printed in China

MIX
Paper from
responsible sources
FSC® C016973

GROW IN THE DARK

How to Choose and Care for Low-Light Houseplants

Lisa Eldred Steinkopf

Photography by HEATHER SAUNDERS

COOL
SPRINGS
PRESS

Introduction

Are you yearning for some green in your life? Have the artificial plants with which you've decorated your home or apartment not fulfilled your need for nurturing a living thing? The only thing they are doing is collecting dust. Adopting a puppy, a kitty, or even a fish may not be possible at this time in your life, but you would love to have something to take care of and "talk" to.

In this fast-paced, stressful world we live in, it may not always be feasible for everyone to take a walk in the park where green living things are in abundance. If you live in a dark, shadowy apartment or an urban area where tall buildings block out your light, it may be even more challenging. Your workplace is probably also devoid of any natural elements. Why not bring nature home to your space with a green friend or two?

Not only do houseplants add beauty to your home, but most importantly, they take things away. It has been proven by NASA scientist Dr. B. C. Wolverton that plants can remove harmful chemicals from our air, called volatile organic chemicals (VOCs). These chemicals come from things that are in all our homes: paint, furniture, carpeting, and electronics. The good news is, one houseplant placed every 100 square feet will remove those chemicals.

Plants also help lower blood pressure and add to one's happiness. If you have heard this good news, I'm sure you have rushed out to buy a plant, confident that it is going to clean the air, lower your blood pressure, and make yourself and those around you happy—only to find that a few short weeks later, not only are you not happier, your plant isn't either and is,

in fact, dying a slow death. How can this barely living plant clean the polluted air or lift anyone's mood? You're frustrated that you'll never have a green thumb, not even light green.

I'm here to tell you that you *can* have a green thumb! It takes nothing more than learning to pay attention to the needs of your plants and meeting those needs in a timely manner. The primary factor to consider when choosing a plant is light. It will determine whether your plant will barely survive or thrive. Quite often, high-light plants are purchased and placed in a low-light situation and vice versa. A perfect example is the succulent craze that has swept the country—most people, especially those in northern areas, simply do not have the amount of light that succulents need to live and grow well.

This book will cover the care of a plethora of plants that can live in a low- to medium-light environment. It will also give you some ideas of ways to not only make the most of the light you have but also improve and supplement your light levels with a simple electric light set-up.

It is not as difficult as you think to find a plant that will live in your light levels. The key is to do a little research before you purchase a plant. So read on, and let's find the perfect plant for your home!

7

CHAPTER 1

Illumination

Illumination is the key to growing any houseplant successfully. Light in any form, be it natural or electric, is the sustenance for your plants. The only time they "eat" is when light is falling on their leaves. When a plant takes in light, water, and carbon dioxide, it converts it to food for itself. This process takes place in the green, chlorophyll-filled cells of the plant and is called photosynthesis. The best part for you, me, and all life on earth is that the byproduct of photosynthesis is oxygen. If it weren't for plants, life on earth as we know it would not exist.

SUNLIGHT

Let's first talk about the light everyone has streaming into the windows of their homes: sunlight.

SOME HOMES RECEIVE more light than others—this depends on the style of home and the number of windows it has.

If light is the key to plant health, how do you know whether you have enough to sustain plant life? If there is enough light in your room to read a book, you have enough to sustain a low-light plant. You may have more light than you realize and can grow a large variety of plants. Initially, you need to determine the amount of light in your home and from which direction it originates. It is important to know whether your windows face east, west, north, or south. Most homes and large apartments have windows facing multiple exposures. If, however, you live in a small apartment, you may have windows that face only one direction.

If you aren't compass-savvy, it's easy to determine an east- or west-facing window based on whether you notice if the sun rises or sets in it, respectively. If you never notice any direct sun at all, you probably have a northern exposure. On the other hand, if the sun shines in your window all afternoon, your window faces south.

Skylights are considered a fifth exposure, as they allow extra light into a room to supplement the wall windows you already have. Lucky you if you have one of these! Let's discuss the five directional exposures and the plants they will sustain.

East

An east window is one of the best for plants. East windows receive soft, cool morning light. Because the sun is coming up over the horizon, it shines in at an angle and reaches quite far into the room. As the sun rises higher in the sky, the area of light in the room will become smaller. African violets, ferns, begonias, prayer plants, aglaonemas, and many other varieties will do well in this exposure. With an east window, the plants you place farther away from it should be those that prefer low light. The windowsill of an east window will sustain medium- and possibly even high-light plants.

West

A west window's light is almost the same as that from one that faces east, but the western sun exposes your plants to a higher level of heat. High-light plants will do well—thrive, in fact—in the windowsill of a west-exposure window. Western light is also on an angle as the sun is setting, so both east and west exposures can support more plants farther into the room. If a plant likes a medium to low light, do not place it on the windowsill of a west window, as that may provide a bit too much light and heat for it. Plants that will thrive in western exposure include cacti and other succulents, air plants, snake plants, ficus, and many flowering plants. Because of the low angle of the sun late in the day, aglaonema, spathiphyllum, and other low-light plants can live 4 to 5 feet back from the window and do well.

⁂ An east exposure receives the soft morning light. Ferns, spider plants, pothos, African violets, and begonias grow well in this exposure.

❦ A west exposure can support cacti and other succulents, as well as air plants. Because the sun sets on an angle, low-light plants, such as aglaonema, can survive further back in the room. If you have a southern or high-light window but would prefer to grow low-light plants, a sheer curtain or blind can cut down the light to an acceptable level to make low-light plants happy.

South

South-facing windows receive the most light throughout the day—it is an intense light that cacti and most other succulents appreciate. In the summer months, the sun is high in the sky, and the sunlight shines directly down into the window, so it doesn't reach far into the room at all. If you hang a sheer curtain or place your plants a few feet back from a southern window, many lower-light plants will thrive in this exposure. If, on the other hand, a low-light plant is placed too close to a south-facing window without protection, it may burn and, if left there, die. The sun is lower in the sky in the winter, so at that time, the light from the sun comes in at an angle. The light from east and west exposures does not change as much in the winter as it does from the south. Low- to medium-light plants may actually do better in a south window in the winter months.

North

North windows never receive any direct sunlight. The plants they can support generally are only foliage plants, such as the cast-iron plant (aspidistra), philodendron varieties, the ZZ plant, and pothos. These plants naturally grow on the rainforest floor, where they live in the shade with only dappled light to sustain them. Flowering plants are usually not an option for northern exposures unless some sort of electric light is added to supplement the sunlight (see page 17). If you have a table lamp regularly turned on in the evening, the small amount of extra light may allow you to grow a higher-light plant than would normally grow in a north exposure, and you may even get an African violet or other flowering plant to bloom.

The Fifth Exposure

Skylights are considered a fifth exposure. They bring extra light into the room directly from above, moving across the room with the sun, and may allow you to grow a greater variety of plants. Be aware, though, that wherever the skylight shines, the light will be an intense light. Though it might not be on your plant for a long amount of time, it may burn low-light plants.

Garden-Level Apartments

If you live in a garden-level apartment, you may receive less light than the apartments above you. All is not lost. If your windows are placed high in your wall, near the ceiling, hanging plants are a good choice. You can also hang a shelf below the window to act as a deep windowsill on which to place your plants, giving you the additional benefit of some extra privacy without having to shut the blinds during the day. Purchase a tree-form houseplant, such as a tall dracaena, whose leaves reach window height. If some amount of light is getting into your apartment, there should be a plant that will survive there. Adding some small grow lights to supplement your light would be helpful too.

🌿 Many plants do well in the low light of a north window. From left to right: ZZ plant, crocodile fern, pothos, and an aglaonema on the floor.

Light Factors

You may be excited to discover that you have a southern-exposure window, yet the light coming in isn't as bright as you hoped and the high-light cactus that should be doing well is struggling. Have you noticed the large overhang extending over your window? How about the awnings that were put up to block the light from fading the carpeting or drapes? What about that enormous tree right outside the window? All these factors determine the amount of light that comes through a window. If you have a deciduous tree right outside—one that drops its leaves in the fall—you will, in fact, have more light in the winter months than you do in the summer. If it is an evergreen, like a Christmas tree, you will have shade the entire year. And what about that huge building next door, blocking all your sun? If it is painted a dark color, it will absorb light; if it's a light color, it will reflect some of the light into your windows. All these factors determine the amount of light coming into your room and the amount falling on your plants' leaves. Read about some creative ways to bring more light into your home on page 20.

❦ This prayer plant is in a low-light area not directly in front of a window. The mirror reflects the light from the nearby window onto the plant, giving it a bit more light.

STEER CLEAR OF VARIEGATED PLANTS

I love variegated plants, or plants with more than one color foliage. They may come in white and green, or green and yellow—or even, as in the case of a croton, red, yellow, green, and orange all on the same plant. The problem with these plants is that they need more light than monochromatic green plants. So, if you only have a modicum of light, steer clear of variegated plants at the garden center, even if you love them. Without enough light, the variegation will fade out of the leaves, and you will be left with a green plant anyway.

⟫ A plant that isn't regularly turned will lean toward the light and grow lopsided. Give your plant a quarter turn every time you water, so it will grow symmetrically.

Phototropism

If you find your plants are leaning toward the window, it is probable that your plant is suffering from phototropism. That may sound like a disease, but it isn't. Sunlight only comes in from one direction in most windows, and, of course, your plant will grow toward the light. This problem can be fixed by simply turning your plants a quarter turn every time you water them. If you notice your plant is bending noticeably between waterings, it may need to be turned more often. If your plant is too large to turn often, place it on a saucer with wheels to make it easier to move. After turning the plant, it will usually straighten out, but if it is a plant with a woody stem, such as a ficus, it may have a small curve in the stem forever—thus, it is imperative to turn your plant often to prevent this.

Plant Tags

When determining whether you have enough light for the plant you have chosen, check its container for a grower's tag and take the time to read it. Many tags are generic and may label something simply as a tropical plant or a houseplant, which only tells you it won't live outside in northern areas. On the other hand, a tag that identifies the specific plant, such as African violet or staghorn fern, will have more specific information

pertaining to it. Try Googling the name on the tag if there is no identifying picture to be certain that the plant listed on the tag is in fact the plant you are purchasing; tags can become mixed up accidentally at the growing facility or garden center.

A plant tag may describe the light needed as high, bright, medium, indirect, part shade, and so on. What is the difference between these descriptors, and what do they all mean? In some cases, they are just different ways of saying the same thing. The chart below decodes some common light descriptors found on plant tags.

LIGHT TYPE	WINDOW
High, bright, direct, full sun	South or west
Medium, indirect, part shade, shade	East; a few feet back from a south or west window
Low light, part shade, shade	North; a few feet back from an east or west window; many feet back from a south window

If, after placing your plant in the light that the tag recommends, you find that your plant is not doing well, it may be that the light level isn't right for the plant. How can you tell whether it is receiving too much or too little light? Read on.

⇛ This low-light plant has been placed outside in too much sun, and as a result the leaves are burnt. If you're moving outdoor plants inside, do so slowly, by first placing them in the shade. Most low-light plants would prefer not to be placed in the full sun outside.

Too Much or Too Little?

Our plants can suffer if they have too little or too much light. What are the indicators you should look for? Let's talk about too little light first. The first indicator is phototropism, as we discussed on page 15. The second sign is that the new growth will be pale with small leaves. In the case of cacti or other succulents, the plants may stretch toward the light and lose their shape. If you have a plant that should normally bloom, such as an African violet or orchid, and it hasn't bloomed within a year, it may not have sufficient light. If you have a plant that collapses for no apparent reason, check the root system. It may be soggy and mushy because the plant didn't have enough light for the roots to use all the water they were given. Plants use more water in higher light. More about that in the next chapter.

You might not have thought a plant could have too much light, but many plants prefer a medium to low light, and light that is too direct or focused on the plant for too much of the day will have adverse effects on them. If a plant gets excessive light or heat, it may wilt severely. If you see this, move the plant a few feet from the window; hopefully, it will recover. Sometimes plants will react to excessive light by curling their leaves down around the pot, trying to draw away from the light, like a vampire. If a plant has been in a low-light situation and is moved to high light, the result will be sunburned leaves. These leaves will not recover and return to their former color or fade to a nice tan, and the growth may be compact and stunted from too much light.

Both too little and too much light can result in the death of a plant if the situation isn't rectified in a timely manner. Pay attention to your plants, and with close observation you'll be able to understand what they are trying to tell you.

ELECTRIC LIGHTS

If you find you do not have enough sunlight to sustain the plants you want to grow, you always have the choice to add supplemental light in the form of electric lights.

EVEN IF YOU LIVE in a room without any windows, or without those that let in enough light to grow plants, you can grow numerous plants with electric lights. There are many options, ranging from small and simple devices to larger hanging lights that blend in better with your décor.

Fluorescent light is the most common type of light people use to supplement their plants. There are different types of fluorescent lights to choose from, including T-5, T-8, and T-12 bulbs, the latter being the oldest variety and the most inefficient type. Many growers use T-5s or T-8s because they are more energy efficient.

I installed simple, inexpensive, 18-inch-long fluorescent plant lights under my cupboards so that I can have African violets blooming almost constantly on my countertop, right next to the coffee maker. It doesn't get much better than that. Many African violet growers use lights to ensure their plants are growing symmetrically and blooming almost constantly; cacti and succulent growers use them to ensure their plants do not etiolate, thus ruining the natural growth pattern of the species they are growing. It's common for people to grow plants exclusively under lights to give the plants the exact light levels and habitat they need to grow to their fullest potential.

LED lights are the newest lighting method used for plants; they are much more energy-efficient than fluorescent lights. While these lighting fixtures are relatively expensive compared to fluorescent lights, they do last much longer and use less energy, as they do not have to be on as long to affect your plants. Many newer kitchens have LED lights installed under the cupboards as part of the design, and there are grow lights that can clip on to any shelf or even your computer screen to allow you to grow plants essentially anywhere you want them.

Incandescent bulbs are too hot and do not supply the complete light spectrum a plant needs to grow properly. Yet, if a lamp is left on near a plant during the evening hours, the extra light provided will still help your plant grow better.

It adds interest in the evening to up-light a plant with a spotlight on the floor, casting cool shadows on the ceiling, but that really doesn't help your plant grow. The chlorophyll, where the energy is manufactured, resides on the upper surface of the leaves. Lighting a plant from above, on the other hand, will assist it in photosynthesis.

In this book, we will be mainly discussing the plants that can be grown with a modicum of light. If you have that amazing southern sun streaming in and are growing cacti and other succulents successfully, you may think you do not need this book. But if you want to grow plants on your coffee table across the room, this book will help you choose the plants that will be able to thrive in a lower-light situation.

❀ Don't discount the light coming from an ordinary table lamp in the evening hours. It can give your plant the little extra boost it may need to thrive.

❀ If you have a low-light spot where you would like a plant, try using a lamp with a specialized grow light. Inexpensive grow light setups can be made with simple grow bulbs and a basic workshop clamp light.

15
WAYS

TO ENHANCE YOUR SUNLIGHT

Here are several ways you can enhance the light you have without installing electric grow lights.

1 Wash your windows! Do you know how much grime and crud build up on your windows? Smog, rain, and dirt in the air settle on the glass, reducing the amount of light that gets through. This is especially true if you live in a rural area with unpaved roads; near farmers' fields being plowed, planted, and harvested; or in an area with high levels of smog. You need to clean your windows more than once a year, and it's especially important to wash them in the fall if you have had your plants outside for the summer. They need all the light they can get when you move them back into your low-light home.

2 Wipe down your plants. A sponge and plain water are all you need to clean away the dust, pet hair, and dirt that build up on your plants' leaves and reduce the amount of light reaching the plant cells. If possible, move them to the sink or bathtub and give them a nice soft shower. When you get that dirt off, your plants will thank you.

3 Remove your curtains and blinds. This may not be advisable in your bedrooms or bathrooms, but if you fill your windows with enough plants, like I do, you won't need any other window coverings. At the very least, keep your window coverings clean so as much light as possible can get through. Remember to raise your blinds during the day so your plants are exposed to the most light possible.

4 If your windows have had a tinting material or privacy contact paper added to them, remove it.

5 Remove window awnings. They block a lot of the light coming in the windows, which, of course, is what they are supposed to do. When you fill your windows with plants, they keep the light from fading the carpeting, so you don't need the awnings.

6 Paint your walls a light color. Lighter colors are more reflective, which means they will bounce more light onto the plants. (Remember to ask permission if you are in a rental property.) That rules out black, navy blue, dark purple, and similar colors.

7 Place mirrors across the room from your windows. It makes the room appear larger and gives your plants a bit of extra light. If you can, cover the wall in cool antique mirrors for plenty of reflected light.

8 Wash your screens. Screens collect dust, pollen, and other wind-blown detritus, all of which can block light. If possible, remove your screens in the fall and keep them out until the spring, as they can cut down the light coming through by as much as 30 percent even when clean.

9 If you have trees outside your home, I'm not going to tell you to cut them down. Trimming them is another story, though. You can selectively prune trees to allow more light into your home and garden. Make sure to find a reputable company to trim your trees and shrubs.

10 If you are replacing or installing a driveway, choose cement over asphalt, as the cement is a lighter color (reflective) and will bounce light into your windows.

11 If you are doing landscaping around your home, choose shrubs that stay shorter over time. I haven't replaced the short shrubs in my front landscape, as I'm trying to find plants that will stay low enough to allow as much light as possible to shine in on my plants.

12 When you see your neighbors outside with paint swatches, discussing the color they are going to paint their house, make sure you put your two cents in and vote for white or some other light color. The reflection off the walls will bring more light into your space for the benefit of your plants.

13 If you need new windows, consider installing larger windows, a skylight, or some other specialty window, which will allow more sunlight to reach your plants.

14 Do not install or hang a stained-glass window or a privacy window where you want to grow plants.

15 If you live in an area where snow is part of your winter, embrace it! That white snow blanketing the ground gives off plenty of light, which bounces into your windows. It is priceless in the dark days of winter.

CHAPTER 2

Hydration and Vitamins

Once you have found the place with the right illumination for your plant to thrive, it is time to decide when and how much to water your plant. Watering too much or too little is the number-one killer of plants, and deciding at what point a plant needs water seems to baffle many. Prevalent also is the misconception that giving your plants fertilizer will solve every problem. While it is an important factor to consider when caring for your plants, blue powder isn't the miracle cure it has been touted to be. Let's dive in and dispel the myths surrounding these two important factors to raising a healthy plant. It isn't as complicated as you think to determine when and how to water and fertilize your houseplants.

HYDRATING YOUR PLANT

A problem that arises when deciding when and how to water is that the labels that come with plants are often misunderstood.

BEFORE FOLLOWING THE label, consider factors that influence how the plant takes up water. For example, the label may advise you to give your plant a cup of water every week, but what those directions do not consider is the weather—maybe your area is going through a cold or cloudy spell, and your plant hasn't been thirsty lately. Or, if you placed the plant in a space with less light than it needs, it may need less water than what the label calls for. Use the label only as a starting point and not as a set-in-stone rule.

Instead of watering your plant on a schedule, as is often recommended, I suggest you *check* your plant on a schedule. When you use an air conditioner in the summer or heat in the fall, it lowers the humidity in your home, drying out your plant faster; during these times, your plant may need more water for a short time while it adjusts to the new environmental conditions. As noted above, a week of cloudy days and cooler weather will decrease the amount of water a plant uses at that time. Conversely, a week of sunny, hot days will increase the water your plant will use. Consider these factors as you check your plant and decide whether to add water.

When to Water

There are a few different ways to check for your plant's water needs. Many people swear by a moisture meter, which has a probe that is inserted into the potting medium and sends a reading to the moisture indicator, letting you know whether it is wet or dry. However, sometimes the reading can be incorrect, affected by the salt content of the potting medium usually present from fertilizer residue. An alternative is lifting your plant's container after watering to get a feel for its weight after it is well hydrated. When you check it the next week, lift it again to see if you can feel a difference in the weight of the plant. If it feels significantly lighter, it is most likely time to water again. If the weight feels the same or a little less, do not add any water.

I can often tell if a plant needs water just by looking at it. Plants that need water are usually a lighter green than their normal color—this is especially true of ferns. This way of reading your plants comes only after working with plants for a long time.

If you are uncertain how to decide if your plant needs water, lift the container after watering your plant to feel its weight. When you lift it again and it feels much lighter, you will know it needs water.

If you find your plant is wilted, it may be a sign the plant is dry. However, check the potting medium before adding water, as wilting can also indicate your plant is too wet. If a plant has been allowed to dry and wilt, it may have damaged the roots to the point that they have died off and may no longer be able to take up water, so the plant wilts just as if the roots had rotted from overwatering. You may be tempted to water the plant after seeing it react this way, but adding more water will not help the plant at this point—the roots that died from drying out will start rotting in the now-overly saturated medium, and your plant will most likely not recover.

If you remove your plant from the pot and discover that it has been overwatered and the roots are black and mushy—and, I warn you, probably extremely smelly—is there hope? If your plant still seems to have life in its green parts, there may be a chance to save the plant. Wash the remaining potting medium off the roots and see if any healthy roots remain. Cut the mushy, dead roots off and replant your plant in fresh potting medium. Many plants will recover from an occasional wilting, reacting only by losing a leaf or two, but this method should not be used as an indicator that your plant needs water on a regular basis.

The best way to decide whether a plant needs water is simply to stick your finger into the potting medium. If you feel moisture at your first or second knuckle, it is not necessary to add water. If it is dry, give it a drink. That said, when you have a plant in a large, deep pot, checking just the top couple of inches may not be a good indicator of its moisture level, as the potting medium may be dry on top but still quite moist in the lower part of the container. Check the potting medium further down in the pot with a long dowel or stick. Push it into the medium as far as you can and hold it there shortly, similar to sticking a toothpick into a cake to check

for unbaked batter; if there is moisture or wet potting medium on the tip of the stick, hold off watering. Water when the dowel is barely moist and never let it completely dry out.

How Much to Water

You've established that your plant needs water, but how much? If it is a cactus or other succulent, you just need to give it a thimbleful, right? I used to think so, fearing that I would overwater it. Or, if the plant is a big drinker, you can leave it standing in water, right? The answer to both of those questions is no. You should give water to any plant, no matter the variety, the same way: until water runs out the bottom of the pot from the drainage hole.

The key to this watering practice is the amount of time that passes before you water again. A cactus or other succulent may not need water again for months, yet a fern or peace lily may need water again in a few days. A plant should stand in water in the saucer for no more than 30 minutes to make sure it has absorbed as much water as it needs. If there is any left in the saucer at that time, dump it out. If your plant is too large to move, use a turkey baster to remove any excess moisture from the saucer.

Top Watering

Top watering is the most common method for watering plants, and it simply means pouring water onto the potting medium at the top of the container. When top-watering your plant, make sure to water all the way around the container, rather than in the same spot each time—this is especially important in large containers. Water needs to reach all of the plant's roots; by watering in only one small area, roots in other parts of the pot may dry out too much and die. Your plant may suffer even though you feel you've been adequately watering, because water was not delivered to every part of the plant's root system.

<<< This scindapsus is showing signs of wilting from being too dry, as its leaves are starting to curl. After being watered, the leaves should straighten out.

>>> The drooping leaves of this monstera are a good indicator that the plant is in need of water. After being watered, the leaves should perk back up.

CACHEPOTS

If you choose to buy a container without a hole, or if your container is an antique or too valuable to drill a hole in, it is better to use it as a cachepot (which comes from the French for "hide a pot") instead. This is simply a container without a hole that your place your plant in, still residing in its utilitarian grower's pot. Take the plant out to water it, let it drain, and then return it to the cachepot. This eliminates any problems that might occur if your plant was left standing in water.

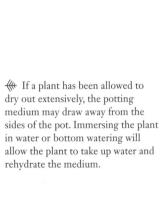

⫷ If a plant has been allowed to dry out extensively, the potting medium may draw away from the sides of the pot. Immersing the plant in water or bottom watering will allow the plant to take up water and rehydrate the medium.

Bottom Watering

Some people like to water their plants from the bottom by adding water to the saucer and letting the plant soak it up. You'll know this process is complete when the top of the potting medium becomes moist; if it doesn't, add more water to the saucer until it does. This method guarantees all areas of the container are hydrated because you can see the moisture reach all the way through the potting medium. After the plant has taken up all the water it needs, remove the excess. Do not leave water standing in the saucer for an extended length of time, assuming it will eventually soak it all up into the root system.

Bottom watering works well, but if you add fertilizer to the water, eventually there will be a buildup of fertilizer salts, which can be harmful to the plant. Once a month, flush the soil from the top to remove the excess salts. Run water through the plant, as you would when top watering, allowing the excess to run out the drainage holes and flushing out any unwanted substances.

Immersion

Quite often the potting medium used by growers when producing houseplants consists mainly of peat moss, which pulls away from the sides of the container when the potting medium dries out completely. When this happens, water runs down the sides of the pot outside the potting medium instead of soaking into it. The water will then drain out the bottom of the pot, not wetting the potting medium at all. To rewet the medium, immerse it in a container of water. You may have to weigh plastic pots down with something heavy or else they will float. Immersion watering should cause the potting medium to swell and fill the pot again. If it does not expand back to the edge of the container, it may be time to repot the plant using a different potting medium or add some medium to the fill the gap between the original medium and the container.

Vacation Watering

If you travel from home for a week or more at a time, it may be necessary to find a way to keep your plants watered while you are gone. If you have a friend or family member you can trust to follow directions and take care of your plants, that is usually the best choice. Or you could get creative: try placing your plant next to a sink full of water and running a string or shoelace from the sink into the potting medium of your plant—the water will wick up the string into your plant, keeping it moist. You could also cover your plants with a clear plastic bag, such as a dry-cleaning sleeve, to keep the humidity high and help your plants stay hydrated. Use dowels or sticks to hold the plastic up and away from the foliage, like a small greenhouse. Move your plants back from the window or light source while you are gone, so they are receiving less light, thus using less water.

Every plant is unique and has different water needs. Consider the weather conditions and time of year, as well as any other surroundings that may affect how much water your plant needs at that time.

Keeping your plant well hydrated, neither under- nor overwatered, can often seem daunting. But if you remember to check your plant often and pay attention to the signs it is presenting, you will find watering is an easy and enjoyable task.

TRACK YOUR WATERING

With our busy lives, it may be hard to remember when you watered and fertilized each plant. I know it is for me. I write it down on my calendar; you could also keep track of your plants in a plant journal or bullet journal. I have plant-themed washi tape and add my own sketches of my plants in my journal. I have also learned to keep track of the dates I repot or up-pot my plants by writing that information on a plastic plant tag that I place in the plant container. Use a pencil, as most other types of pens or markers eventually wear off.

When you go on vacation, wick-watering your plant is a good way to keep it hydrated. Insert one end of a string or shoelace into the potting medium and place the other end into a sink of water. The water will wick up the string into the medium, keeping it moist.

FERTILIZING YOUR PLANT (A SHOT IN THE ARM)

Light is the food source for your plants.

MANY ASSUME THAT when a plant is given fertilizer, it is being "fed," but I view fertilizing my plants like taking a daily vitamin—but unlike humans, plants do not need fertilizer on a daily basis.

Often it is recommended to fertilize your plants once a month. That suggestion goes along with the belief that you should also water your plants once a week, whether or not they need it. The proper rule of thumb is to fertilize your plants every fourth watering, thus the once-a-month ideology. In reality, the fourth watering may occur more or less frequently than once a month, depending on the plant, its container, whether the roots fill the container, the weather, and time of year.

I suggest that, instead of giving your plant full-strength fertilizer every fourth watering, you use a ⅛- to ¼-strength formula every time you water. That way your plant is benefitting from a steady supply of nutrients instead of one large dose all at once.

Which fertilizer you use is completely up to you. When you purchase fertilizer, you will notice three numbers displayed prominently on the container. These indicate which macronutrients the fertilizer contains and in what percentages. Some gardeners think of these numbers as representing "up, down, and all around." The first represents nitrogen (N), the nutrient beneficial to the green, or "up," part of the plant; it assists your plant when putting out new growth. The middle number represents phosphorus (P),

🌿 There are many forms and kinds of fertilizers to choose from, including organic and nonorganic types. Clockwise from the top: liquid fish emulsion, slow-release granules, water-soluble blue crystals, and slow-release sticks.

which helps the plant form strong roots ("down") and helps flowering plants grow brighter, bigger, and longer-lasting flowers. Fertilizer does not make the plant bloom, though—the only thing that does this is the correct amount of light. The last number represents potassium (K), which is beneficial for the "all-around" health of the plant. It helps the plant resist disease and assists it with drought and cold tolerance. A container of fertilizer with the numbers 20-20-20, for example, is made up of 20 percent nitrogen, 20 percent phosphorus, and 20 percent potassium. The other 40 percent of fertilizer consists of smaller amounts of micronutrients and filler of some kind that holds the formula together in a digestible format for the plant.

Types of Fertilizer

There are two types of fertilizer to choose from— organic and nonorganic (synthetic). Both have their pros and cons, and both are beneficial to your plant's health when applied properly.

We'll discuss the nonorganic forms first. These deliver nutrients to the plant quickly, but if too much is applied, it will burn leaves and can affect the entire plant. Probably the best-known nonorganic fertilizer is the synthetic blue crystal form that is measured in a scoop and then poured in water, where it dissolves and turns the water blue. This is called water-soluble fertilizer. It is easy to use and find, which makes it a popular option.

Nonorganic fertilizer also comes in concentrated liquids that can be added to water to produce the right strength and granular forms that are applied to the potting medium and dissolve when water is added to it. Another type of fertilizer is most likely already in your plant when you buy it—a slow-release fertilizer in the form of small, round beads. They are colored (most commonly blue, green, or cream) so they are easy to see when applied by the plant growers. These small beads

If your leaf is pale with dark veins, as in this picture, or is dark with pale veins, it may need to be fertilized, as it has a nutrient deficiency.

are encapsulated in a coating that dissolves slowly when the plant is watered, releasing a small amount of fertilizer. These are also called continuous-feed fertilizers, as they last for up to three months in the potting medium, continuously releasing fertilizer in small amounts to your plants. If you buy a plant and can see the fertilizer beads, it may not be necessary to add fertilizer for a few months.

Slow-release fertilizer also comes in small stick forms for houseplants. These are pushed into the potting medium at equal intervals around the container edge, the size of the pot determining how many sticks are used. However, as this concentrates the fertilizer in one area, it may damage nearby roots and not reach the entire rootball.

The other type of fertilizer is organic, meaning it is derived from the remains of living organisms. Organic fertilizers release nutrients slowly to the plants. Fish emulsion is a popular form that may be combined with kelp meal, blood meal, and worm castings. Organic fertilizers break down slowly and are less likely to burn your plants if they are overapplied or if residue is left on the surface of a leaf.

You may try a few fertilizers before you find the one that works best for your plants. There are also fertilizers for specific plants, such as for cacti and other succulents and for African violets. These aren't necessary to use for those plants but are formulated to have the correct percentage of each nutrient for that particular plant family, so they eliminate some of the guesswork in choosing a fertilizer.

When & How Much to Fertilize

How often should you fertilize? As mentioned on page 31, you can do so every fourth watering using a full-strength formula or every time you water using a ⅛- to ¼-strength formula. However, because houseplants grow indoors, I don't recommend using full-strength fertilizer. Never assume that more is better and apply more fertilizer than the package calls for.

You should also be fertilizing only during the time of year your plant is actively growing—from approximately March through September, depending on the part of the country you live in. As soon as you see new growth on your plant in the spring,

begin your fertilizer regime. When the days get shorter in the fall, forgo fertilizing until spring. Plant growth slows down when the days are shorter, so they do not need added nutrients and may not use them if you add fertilizer.

Additionally, too much fertilizer applied to your plant can severely damage or possibly kill it. Usually, overfertilizing burns the plant, meaning parts of it may turn black and die. If only a section of an overfertilized plant dies, you may be able to flush the potting medium with water to wash the excess fertilizer out. Hopefully, your plant will recover.

If you have an ailing plant, it is tempting to reach for the fertilizer, thinking you are helping it, but a weak plant should not be fertilized. Instead, find out what the problem is and decide on a course of action. If your plant has many yellow leaves or spots on the leaves, or if it seems to be losing vigor, it may be time to reconsider the location where you have it placed. It may need more light or a different watering regime. It may have fungus or disease that needs to be treated. Take a picture or a leaf of the plant to a garden center and the experts there should be able to help you determine what your plant needs. Remove any yellow leaves—they will not turn green again.

If your plant's leaves are all a lighter green than they should be with darker green veins, or darker green with light veins, your plant may indeed need a boost of fertilizer, as this indicates it is missing nutrients it needs to keep its leaves green.

The key to keeping your plants healthy and looking good is paying attention to them. If they are left to languish in a corner, treated simply as part of the décor rather than a living thing, don't be surprised when problems arise. Plastic plants may be better in those spots if you want something there just for the aesthetic. Give your plants the correct light, water, and fertilizer and you will be amazed at how good they look!

CLIMATE & ATMOSPHERE

Our plants come to live with us after leaving growing facilities that have near-perfect conditions.

THE ONLY JOB the plant nursery had was to keep the plant as healthy and beautiful as possible for retail sale, giving it the optimum amount of water, fertilizer, and care. Then the plants are shipped north and are brought to our homes, where we attempt to take care of them to the best of our ability as well. Still, we don't have the optimum conditions they want. Let's talk about some of those conditions and how we can improve them.

Temperature

Plants are usually grown in climates much warmer than we are accustomed to. Indoors, however, they are usually comfortable in the temperatures we are comfortable in—between 60° and 75°F. If you turn your thermometer down at night, you are making your plants comfortable as well; they prefer a drop in temperature at night of approximately 10°F.

Humidity

Our homes in the winter can feel like the Sahara Desert to both our skin and our plants. Most houseplants hail from the tropical regions of the world, where humidity levels are as much as 80 to 90 percent. In our heated homes in the winter, however, humidity could be as low as 20 percent. Plants do better when they have sufficient humidity. How can we accomplish this?

Many people turn to a handheld mister, yet a mister only temporarily raises the humidity. Misters can also leave excess water on the leaves, which can contribute to the conditions needed for a disease to begin. Adding a humidifier to your furnace or plant area would be ideal. If that isn't possible, group your plants together to raise the humidity around them.

If you have only a few plants that wouldn't do well in the same light situation, you can raise the humidity by placing your plants on a pebble tray. Use a saucer larger than the one your plant is standing on and fill it with small pebbles. Add water to the saucer so that it just comes to the top of the pebbles, then set your plant saucer and pot on top. You can set the pot directly on the pebbles as long as the plant is not standing in water. As the water evaporates from the pebbles, it will rise into the air around the plant, adding humidity. Keep the pebble saucer filled with water for best results.

Air Circulation

Air circulation may seem like a weird thing to talk about inside your home, but stagnant air is not conducive to your plants' health. Moving air helps keep bugs at bay, strengthens the stems of your plants, and keeps leaves dry, which minimizes the chances of a disease developing. Open a window occasionally or add an electric fan to your plant area. A small amount of moving air will help keep your plant healthy.

The healthiest plants are kept well watered, in humid spaces, and given a shot in the arm on a regular basis. A healthy plant is better able to ward off insects and disease; one that does not have consistent care is more likely to attract mites or insects and develop disease. The goal is to have a plant that is easy to take care of, looks healthy, and adds beauty to your home. Providing the right conditions goes a long way toward ensuring you have a healthy, attractive plant for many years to come.

It can be helpful to have a fan for air circulation, as plants do not appreciate stagnant air. Clippers are invaluable for trimming unruly plants, twine is perfect for tying up leaning plants, and pebble trays are very important when trying to elevate humidity levels.

To keep the humidity high around your plants, place them on pebble trays that are filled with water. Make sure your plants aren't standing in the water, though, to avoid rot.

VINING WINDOW DÉCOR

Is your pothos or heartleaf philodendron getting out of hand? Are the vines trailing to the floor and the cats are batting at them? Why not move your vine to a stand or shelf by the window and frame your window with vines? There a few ways to accomplish this. The easiest way is to just pound nails in the wall and drape the vines over them. Cup hooks may work better and are easily screwed into the walls. If you don't want to make a large number of holes in the wall, or if you are renting and it isn't allowed, you can find adhesive hooks at a local hardware or home store that stick to the wall but come off easily when they need to. What a wonderful way to frame your view out the window!

❦ Adhesive hooks are a perfect solution to help train a plant around a window or up a wall. But be sure the strip is large enough to accommodate the plant's future growth, or else the hooks may harm the plant. Check the plant often to prevent this damage from occurring.

❦ If you have a window you would like to frame with vines, use small nails—or adhesive hooks, if you are renting or don't want to damage your walls—to rest the stems where you want them to climb.

CHAPTER 3

Maintenance

It is exciting to bring home a new plant, like obtaining a pet that doesn't need to be walked in the rain. Once you get it home, it will beautify your space and make every day a bit brighter. As time goes by, grooming and cleaning your plant (and perhaps talking to it) will become a therapeutic part of your day. In this chapter, we will look at the process of buying the perfect plant and making sure it is potted in a suitable container with the potting medium that will ensure it grows to its potential. We'll also look at how to take care of plants via grooming and battling diseases.

HOW TO BUY A HOUSEPLANT

The most important way to maintain a healthy plant is to begin with a healthy plant from the start.

IT IS IMPORTANT to choose a reputable place to buy your houseplant—and by "reputable," I mean one that has a staff dedicated exclusively to caring for the plants. Quite often, chain stores that sell plants care for them as an afterthought, usually when someone notices they are dying of thirst. I recommend going to a houseplant store or to an independent garden center. These businesses usually have a staff member who exclusively cares for their plants and knows about the needs of each.

Think twice before "rescuing" a plant from the clearance area; these are often beyond saving. And don't buy a houseplant that is sitting outside in the full sun, being wind-whipped and sunburned. This is especially important if you are buying a plant for the first time. Choose a plant that is in good health and does not have yellowing leaves or other signs of distress.

Have a spot in mind in your home where you want to place your plants and be sure to know roughly the amount of light the spot receives. As a plant fanatic, I know how easy it is to buy plants on impulse. When I do this, I usually know whether the plant I have selected will survive in my home because I'm familiar with the level of sunlight my windows receive and the space I have available near them. If, on the other hand, you are buying your first plant and have no idea of the sun exposure you have to offer a plant, it may be better to do some research first.

❦ When shopping for a houseplant, shop at a reputable garden center, where there is staff that can assist you with choosing a plant and will be able to inform you about its proper care.

← Even though plants on the clearance rack seem like a good idea because of the price, it is better to begin with a healthy, problem-free plant. Why bring home a plant that is a challenge right from the start?

When shopping for houseplants, how do you find the perfect one? Look for one that is a healthy green color, is standing up in the pot (unless it is a vine, of course), and is full—not sparse—and well watered.

Well watered does not mean it is standing in water in the saucer or pot cover. Make sure to always check in the pot cover, if the plant has one. If it is full of water, keep walking. One does not know how long the plants have been standing in water and whether root damage has already occurred. Well watered means the potting medium is damp to the touch, not bone dry or overly soggy. Also check the leaf tips—are they all brown or even black? This could indicate inconsistent practices that have allowed the plant to dry out too much between waterings. One or two yellow leaves on a good-sized plant probably isn't cause for alarm; the plant is simply adjusting from the near-perfect growing conditions provided in its previous environment and will drop leaves as a normal process of acclimation.

If you see a large number of roots protruding out of the bottom of the pot, this may mean the plant is rootbound and needs a new pot. Look at the back sides of the leaves and in the axils of the leaves (the angle where the leaf meets the stem) to check for any bugs or insect eggs that may be lurking there. Once you find the plant you like and have inspected it to make sure there are no obvious problems, buy it.

When you get to the register to pay for your plant, ask for the plant to be wrapped up, preferably in a paper sleeve or a paper bag, at the very least. This is a must in northern winters, but it should be done even in the fall and spring, if it is a cold day, and even more importantly if it is a windy day. Seeing houseplants leaving the garden center's parking lot in the uncovered bed of a truck makes me cringe. These plants cannot handle the whipping wind or cold temperatures, and this shock may kill an otherwise beautiful, healthy plant. When you have transported your plant safely home, watch it closely for a few weeks to make sure it is adapting to its new habitat and no surprise visitors (pests) have appeared.

Repotting Your Plants

If you buy a plant and it is in a utilitarian, unattractive black or brown grower's pot, you may want to replace it. If you are buying a new plant, you will most likely just be repotting your plant in the same size container, not up-potting it to a larger size (see below). Buy a container that is proportionate to your plant and approximately the same size as the one you are taking it from. Slip the plant out of the grower's pot and

into the new one. Check that the depth of the plant is correct. Quite often, plants are planted too deep and their stems are covered with potting medium. How can you tell it is potted too deep? The point where the roots begin is usually where the medium should start. If the stems are covered with too much medium, it may cause rot. The extra medium needs to be removed and the plant repotted at a better depth.

⋙ When a plant is root-bound, as is the one on the left, tease the roots and up-pot the plant into a container the next size up, which has room for the rootball to expand.

Make sure your plant is placed in the pot at the right depth, leaving approximately ½ to 1 inch of space between the medium and the rim of the pot. This allows you to hydrate the plant without the water and medium running over the sides.

HOW DOES A HOUSEPLANT BECOME A HOUSEPLANT?

I asked Mike Rimland, the plant hunter for Costa Farms, a few questions about how a plant is chosen to be a houseplant and how it gets from the jungle to our windowsills.

How does a plant hunter find a plant?
I travel the world in subtropical and tropical areas, which is where low-light, indoor-possible plants are from. I test the plants myself in real-world indoor environments of home and office. I use a variety of different locations based on light levels and other climatic conditions to ensure a plant will truly be suited to life indoors.

How is a plant tested, and what are you looking for?
I start with an understanding of typical indoor environments and the minimum light-level needs of a plant from a metabolic standpoint. I also take into account the water consumption indoors. This has taken me many decades of trial and error, learning all the factors that must be considered.

How long is it from the time a plant is found until it goes to market?
This can range from four to eight years, based on testing and reproduction times, to build supply

to make plants for sales. With many new plants I've introduced, we start with a small amount. It's a process building up enough inventory to ship to retailers across North America.

How many do you need to have stocked to release a plant in the market?
It depends on the plant and the market we see for it. There is a wide range, based on factors such as the genus and pricing (which determines demand levels). It is also based on the geographic factors of distribution. In general, we look to have a minimum of fifty thousand.

When up-potting your plant, choose a pot that is only one size larger than the pot it was previously in, gradually moving your plant up to a larger size pot as it grows. It's best not to give plants too much extra room, as water could sit in the extra potting medium and cause rot.

Up-Potting Containers

When up-potting, or moving a plant to a larger pot for the plant's health, do it gradually. If you have a 4-inch plant, move it to a 5- to 6-inch pot; a 6-inch plant should be moved to an 8-inch pot; an 8-inch to a 10-inch, and so on. Moving a plant to a much larger pot can cause problems. Too much potting medium around a rootball can lead to root rot, as the roots cannot use all the water held by the extra potting medium.

If your plant shows signs of needing water more often than usual, meaning more than once a week, that is an indicator it may be rootbound and needs a larger home. Take the plant out of its container and examine the roots. Are they filling the pot? Are there more roots than potting medium? Time to up-pot it.

If, when you water your plant, the water sits on top of the potting medium and takes a long time to drain out, it may be that the pot is full of roots or the potting medium is broken down and compacted in the container. That would be an indication that the plant needs either a larger pot or, perhaps, fresh potting medium.

On the other hand, the water may run immediately out of the pot without wetting the potting medium at all. If the potting medium has been allowed to dry out completely, it will shrink away from the sides of the pot and leave a gap between the potting medium and the side of the container. The water then runs down the side of the pot without hydrating the plant. This happens with potting medium that is largely made up of peat moss. It may need a larger pot, or perhaps just a better potting medium. A short-term solution is to submerge the container entirely into water for a short time so that the potting medium can soak up water and re-expand (see page 29).

Before repotting or up-potting a plant, make sure it is well watered. This helps to keep the rootball intact when handling the plant and thus helps the plant experience less shock. If your pot is pliable, gently squeeze it to dislodge the plant. If it is a rigid container, try turning the pot upside down with your hand over the rim; hopefully the plant will fall into your hand. If not, give it a gentle tug, being careful not to damage the stem or the leaves. If this doesn't work,

run a knife around the edge of the pot to loosen the rootball from the pot. If the plant is rootbound, these roots might be tough to remove from the pot. If all else fails, it may be necessary to cut the pot off if it is plastic, or break it entirely if it is terracotta or ceramic, to get the plant out. Luckily, those drastic measures aren't often necessary.

After removing the plant from the pot, check the root system. If your rootball is tangled and densely packed, carefully tease the roots apart. If the roots are discolored and mushy, cut them off. Usually, roots are a nice white or light color and firm to the touch. Cutting slices down the side of a solidly packed rootball will stimulate new roots to grow into the fresh potting medium. As you settle your plant into its new home, leave room between the potting medium and the rim of the pot so that, when watering, potting medium and water do not run over the rim. A small pot should have ½ to 1 inch of room, and a larger pot should have 1½ to 2 inches or more.

The best time to up-pot a plant is in the spring, when the plant starts actively growing. In the northern areas of the country, that is usually late February or early March and can continue into the summer months. It is best not to up-pot your plants as the days get shorter in the fall, when plant growth slows down and almost stops. You can still repot your newly purchased houseplants into new, more attractive containers of the same size at that time, though. If you up-pot the plant in the fall, it will most likely result in root rot because the extra potting medium will hold too much excess water at a time when the plant needs less of it. Up-potting in the fall may be necessary, however, if you have just purchased an extremely rootbound plant from a garden center. If that is the case, be extra cautious with your watering practices through the winter.

When potting a plant after adding extra potting medium, firm it in gently without packing it down too much. Remember, your plant roots need oxygen, and packing the medium in the pot too firmly may collapse the air spaces between the particles. After repotting or up-potting a plant, give it a good drink. Watering it in is the best way to settle the medium.

Drainage

The most important factor for choosing a container is the drainage hole. A drainage hole is a *must*. Trying to keep a plant well taken care of is hard enough without worrying whether it's drowning because there is no place for excess water to go. A drainage hole allows excess water to run down through the pot and out the bottom. Your container should also have a saucer to catch the superfluous water.

There are many types of containers to choose from. The usual materials are terracotta, glazed terracotta, and plastic. Terracotta is porous, allowing water not only to come out of the drainage hole but to transpire through the pot's walls as well. These pots are good choices for plants that do not like to be kept overly moist. Plants that benefit from being in terracotta include cacti and other succulents. Glazed terracotta has been colored and sealed so that it no longer loses water through the pot walls. Another benefit of terracotta and glazed terracotta is the weight of the pots. If you have a plant that is top-heavy, especially when it is on the dry side, the weight of the pot will keep it upright.

Plastic pots are nonporous and thus keep a rootball moist longer than terracotta. Plastic and glazed terracotta are good choices for plants that need to be kept moist, such as ferns and peace lilies. Of course, there are many unusual pots to choose from as well. Almost anything that can hold potting medium will work as a planter. Just remember the drainage hole!

❧ If a container you would like to use for your plant doesn't have a drainage hole, you can easily drill a hole with a diamond-tipped drill bit or masonry bit.

CLEAN YOUR CONTAINERS

If you are using a container you already have at home, or one you've inherited, make sure to clean it before using it for a new plant. If it has been previously used, wash the dirt off and then soak it in a solution of 10 parts water to 1 part bleach. This will eliminate any problems, such as disease, that may be lurking on the pot.

Potting Medium

You may have noticed I use the words "potting medium" instead of "potting soil." Why? Most plants are not grown in a mix that contains soil. It has been replaced with sphagnum peat moss, which is harvested from peat bogs. Because of the concern in the last few years about the mining of peat moss bogs and its effect on global warming, many companies now use coir, a byproduct obtained from the coconut husk industry, which makes floormats, ropes, and more. Both peat moss and coir are known for their water-holding capabilities. These products are usually combined with vermiculite and perlite to help with drainage, keeping the roots from rotting in waterlogged medium. A wetting agent is also added to commercial potting mixes to assist with the rehydration of the peat moss and coir.

When buying a potting medium for houseplants, purchase one without added fertilizer or water-retaining beads. Those products are better suited for outdoor containers. Many potting mixes for houseplants seem too dense for the roots, and the roots cannot get enough air. I suggest adding more vermiculite and perlite to a purchased potting medium, creating a mix that is one-third of each and one-third purchased potting medium. Mix it well and slightly moisten it prior to potting up your plants.

🌿 Often, commercial potting medium is too heavy for many houseplants, and doesn't offer proper drainage—but you can easily amend it. It is a good idea to add ⅓ vermiculite (left) and ⅓ perlite (bottom) to ⅓ purchased potting medium (top) to ensure that the drainage is adequate for your plant.

POTTING MEDIUM LEVELS

When repotting or up-potting a houseplant, place your plant in the container you have chosen, leaving a gap between the potting medium and the rim of the pot. If the potting medium is all the way to the top of the container, the water and potting medium will run over the edge and make a mess when you water the plant. A small pot of 2 to 6 inches in diameter, should have ½ to 1 inch of room and a larger pot 1½ to 2 inches or more. This allows you to add water without it washing over the rim of the pot.

GROOMING

Keeping your plant clean and looking good is as important as keeping ourselves groomed.

WE WASH THE dirt off our bodies daily yet may allow our plants to be dusty and grimy for months or even years at a time. Remember, they are arriving to your home from a spa environment. They have been pampered and groomed every day to look their best for potential buyers. While you likely won't maintain that regimen, keeping them clean can make them feel more at home.

When watering your plants, if possible, move them to a sink or shower and give them a good rinse. This will remove dust and dirt and dislodge any pests that may have dared to take up residence on your green friend. If that is not possible, wet a sponge or paper towel and wipe down the leaves, always washing your sponge or cloth between plants so as not to introduce unwanted guests from one onto another. Keeping the dirt off their leaves also helps the leaves receive the maximum amount of unobstructed light.

You may have been told to use milk, mayonnaise, or a commercial plant shine on the leaves of your plants to clean them and make them extra shiny.

Please do not do this! Using these products clogs the leaves' stomates, which are comparable to the pores of our skin. Food products may also attract insects or our pets to nibble the leaves. Many plants are poisonous, and we certainly to do not want to encourage any living thing to chew the leaves of our plants.

PROBLEM SOLVING

Hopefully, by purchasing your plant at a reputable place and regularly checking it for insects or diseases, you will avoid most problems that could occur over the life of your plant.

YET THERE USUALLY comes a time when at least one problem will arise. Being observant every time you interact with your plant is the best way to catch a problem before it gets out of hand.

Pests

Let's talk about pest problems first. Pests may come into our plants' lives from another plant that may have been brought in, or eggs may have already been on the plant when it came into your home. Often, the eggs are so minute that we can't even see them or may think they are a speck of dirt. Once they hatch, they can reproduce quickly, so, again, it's best to discover them early.

Plants will gather dust over time, so it's important to remember to wipe them clean with a lightly damp sponge. This will help them maximize their light absorption. Do not use plant shine products.

What are you looking for? A common indicator that you have a problem may be a shiny, sticky substance on the leaves or stems—this substance is called **honeydew** and is the excretion of insects that suck the juice out of the plant. If you see this and it doesn't wipe off easily, as water does, you have a problem. In the advanced stages, you may notice a black, sooty mold covering the honeydew, which may be more obvious to the naked eye than the shiny honeydew alone. Ants are also attracted to the honeydew, so the presence of ants is another clue you have a problem. Taking care of the problem (see page XXX) will most likely get rid of the ants as well. There are a few other common houseplant pests you may encounter.

Mealybugs: If you notice a white cottony or lint-like substance on the leaves or in the axils of the plants, you may have something called foliar mealybugs. Mealybugs are slow-moving scale insects that pierce the leaves of the plant and suck the juices out. They can be hard to get rid of, but with consistency, you can bring them under control. The best scenario is that you discover them early on when there are only a few of them. Use a cotton swab dipped in rubbing alcohol and touch each white cottony area to kill the insect. Use a spray of Neem oil or horticultural oil to smother any missed or newly hatched insects. Insecticidal soap, foliar insecticides, or a systemic insecticide may also be used. Completely read the labels of all products used and follow the instructions to the letter for your safety.

Root mealybugs are harder to detect, as they live in the potting medium. If your plant is showing signs of distress and no insects or other problems are evident, pull the plant out of the pot and check for mealybugs—they will resemble pieces of rice stuck to the roots. They also secrete honeydew, and the inside of the pot may be sticky. Use a systemic insecticide to eradicate them.

✤ If you find a white, cottonlike substance on your plants, it may actually be mealybugs, a type of slow-moving scale insect.

Aphids: Sometimes called plant lice, aphids are mostly found on the new growth of your plant. They are easy to see with the naked eye and can be different colors, including black, green, red, and yellow. Like mealybugs, they secrete honeydew. You can easily remove them with a strong stream of water or by wiping them off with a paper towel. A contact or systemic insecticide may also be used if necessary. These practices are best done outside, but if it is too cold, wash the plant in the bathtub. If you decide to use an insecticide, do so in a room that can be closed off from the rest of your home.

Scale: These insects are similar to mealybugs, but instead of a white cottony covering, they have a brown covering and they do not move once they find a place they like. Scale insects are harder to see, as they may look like part of the plant. They can attack any plant and also secrete honeydew. Remember, if you have honeydew, it's because an insect is producing it. Look closely for the offender. If scale is present in small numbers, it can be picked off with a fingernail.

❦ Scale is a small insect protected under a hard shell that sucks the juice out of your plant. If it is left untreated, it may eventually kill your plant.

❦ The shiny, sticky substance on this leaf is evidence that there is an insect problem on your plant. This sticky material is the excretion of the scale insect present on this plant.

If the insects are present in larger numbers, it may be necessary to use a systemic insecticide or to smother them with a Neem or horticultural oil.

Spider mites: If you notice that the leaves of your plant are turning a mottled yellow-brown color, it may indicate that you have spider mites. The mites, which are very small creatures in the spider family, pierce the leaves with their mouth parts and suck the juices out, discoloring the leaves. If they are present in large enough numbers, you may see webbing on your plants and may actually see the mites. They like to attack dry plants or plants in areas where furnaces run continually—the lower humidity is like a dinner bell calling them to supper. Use pebble trays to raise humidity, as described on page 34 and make sure your plants do not dry out. If you have mites, start by washing your plants to remove them; it may also be necessary to use a miticide, as insecticides will not work on mites. A spray of Neem or horticultural oil will be helpful as well.

❦ When spider mites attack your plants, the leaves take on a mottled appearance. The mites have sucked the cell contents out of the leaves, creating pale splotches like the ones seen here.

Fungus gnats: If you ever are bothered by minute black flies flitting about, you may have fungus gnats. Many people mistake them for fruit flies—but if plants are present and rotting fruit is not, there is a good chance you have fungus gnats. The larvae of these small flies live in the top few inches of overly moist potting medium and eventually turn into the adult flies. As the larvae live only in the top 1 to 2 inches of the moist medium, allow your plant to dry down more before watering again. The larvae will not live in the dry medium and the number of adult flies should be greatly reduced, if not eliminated. If the infestation is abundant, remove the top few inches of potting medium and replace it. A systemic insecticide may also be applied.

Whiteflies: As the name suggests, these are white flying insects, but they are not flies at all. You will usually find them on the undersides of leaves, where they feed and lay eggs. They will be apparent before turning the leaf over, however, as they fly when the plant is disturbed—this can make it difficult to get rid of them. Yellow sticky cards made specifically for this can be placed in the pot, as the yellow color is attractive to them, and they'll hopefully stick to the card and die. A spray of Neem or horticultural oil may effectively suffocate them, or you can try applying a systemic insecticide to the potting medium. Sometimes using a vacuum to suck the insects out of the air works to decrease their numbers.

Diseases

If your plant seems to have a problem, but there aren't any insects or mites in sight, it could be suffering from a disease. What should you look for?

Powdery mildew: These fungal spores appear as a white powdery substance on the leaves of a plant and are spread by water or even a breeze. The combination of poor air circulation, low

❦ Powdery mildew is a fungus that can cover the leaves of a plant with a white flourlike substance that, unless controlled, can eventually kill a plant. It prevents the plant from being able to photosynthesize well because the light is blocked.

light, and cool temperatures contribute to the growth of this fungus. If left untreated, it can spread and eventually kill the plant. Keep your growing area well ventilated, your plants spread apart so they aren't crowded, and your plants' leaves dry to avoid powdery mildew. If it does appear, remove the affected plant parts or spray them with a fungicide. Neem oil also works well.

Crown rot: This disease is not usually apparent until your plant collapses, and it will be noticeable in the center of the plant where the leaves rise out of the potting medium (the crown). It occurs when the potting medium has been kept too wet or your plant is planted too deeply in the medium. Crown rot is usually fatal to the plant.

Leaf spots: Bacteria or fungi can cause these spots when water sits too long on the leaves of a plant. If you find it on the edge of a leaf, trim that part off. If it is in the middle of the leaf and is spreading, cut the whole leaf off. It may help to spray a fungicide or Neem oil to prevent the spread of the disease.

Black sooty mold: As mentioned on page 48, this mold grows on the honeydew, or excrement, of certain insects. Eradicate the offenders and then wash the leaves of your plants, removing the sooty mold and the honeydew at the same time.

Environmental Problems

Houseplants may exhibit problems that are not be caused by insects or a disease at all, but rather by the environment they are in or the way they are being cared for.

Cold damage: Our houseplants can suffer cold damage even if the temperatures are above freezing. Many plants can be affected if the temperatures fall below 50°F, though fatal damage doesn't usually occur until the temperatures are in the low 40s and 30s. Cold damage appears on the plant as leaves that have turned brown or yellow and they drop from the plant. Many tropical plants in particular will drop leaves and even die in too-cold temperatures, as they like temperatures between 60° and 80°F. That said, if you are comfortable, your plants probably are too.

Brown leaf tips: Some plants—such as spider plants, prayer plants, and dracaenas—are very sensitive to dry air, fertilizer salt buildup, and damage caused by inconsistent watering practices, and the tips of their leaves will turn brown as a result. Adjust your watering and fertilizing practices and raise the humidity around these plants to rectify the problem. Trim the brown off the leaf tips, following each leaf's true shape so they match the appearance of the healthy leaves.

Wilting: This is usually a sign the plant needs water, but sometimes a plant wilts because it has been overwatered and is suffering from root rot. In the case of root rot, the plant really is dry but can no longer take up water through its rotted roots. Wilting may also be from crown rot, which will cause the plant to collapse. If your plant wilts while the potting medium is dry, water it and your problem is solved. If it wilts while it is moist, then investigate further.

Leaf drop: All plants naturally drop their leaves at some point as the leaves age and fall off. But if the floor is littered with leaves, there is a problem. When you move a plant from a high-light situation to a low-light one, the plant will react by dropping leaves until it has only the number left it can support with the amount of light it is receiving. Plants may drop leaves if they have been over- or underwatered, as well. Consistent, even moisture is best for your plants.

Leaf discoloration: Are some leaves a different color than others? If this happens on new growth, it usually isn't a problem; new growth is often a different color than the older leaves. If the leaves remain off-color or have veins that are a different color than the leaves, however, the plant may be experiencing a nutrient deficiency, or it could be an indication that the pH of the planting medium is too high or too low. If the latter, nutrients cannot be released to the plant. If you test your medium with a pH meter, it should read near a 6.0 to 6.5. If the pH is much higher or lower than that, it may be harming the plant, and repotting the plant in fresh potting medium will most likely solve the problem.

Taking good care of your plants really boils down to paying attention to them and their specific needs, which keeps complications to a minimum. The healthier a plant is, the less likely it will be attacked by pests and diseases.

CHAPTER 4
Plant Profiles

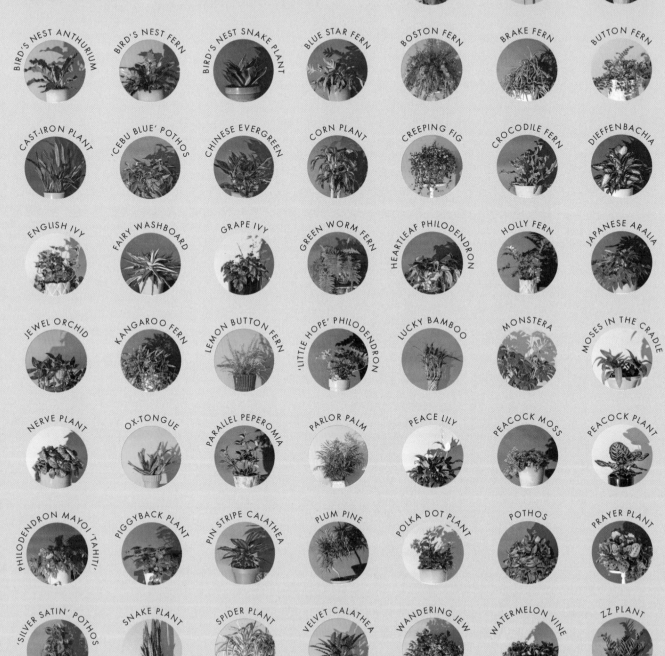

ANGEL VINE

'ANITA' DRACAENA

ARROWHEAD VINE

BIRD'S NEST ANTHURIUM

BIRD'S NEST FERN

BIRD'S NEST SNAKE PLANT

BLUE STAR FERN

BOSTON FERN

BRAKE FERN

BUTTON FERN

CAST-IRON PLANT

'CEBU BLUE' POTHOS

CHINESE EVERGREEN

CORN PLANT

CREEPING FIG

CROCODILE FERN

DIEFFENBACHIA

ENGLISH IVY

FAIRY WASHBOARD

GRAPE IVY

GREEN WORM FERN

HEARTLEAF PHILODENDRON

HOLLY FERN

JAPANESE ARALIA

JEWEL ORCHID

KANGAROO FERN

LEMON BUTTON FERN

'LITTLE HOPE' PHILODENDRON

LUCKY BAMBOO

MONSTERA

MOSES IN THE CRADLE

NERVE PLANT

OX-TONGUE

PARALLEL PEPEROMIA

PARLOR PALM

PEACE LILY

PEACOCK MOSS

PEACOCK PLANT

PHILODENDRON MAYOI 'TAHITI'

PIGGYBACK PLANT

PIN STRIPE CALATHEA

PLUM PINE

POLKA DOT PLANT

POTHOS

PRAYER PLANT

'SILVER SATIN' POTHOS

SNAKE PLANT

SPIDER PLANT

VELVET CALATHEA

WANDERING JEW

WATERMELON VINE

ZZ PLANT

If you are living in a low-light situation, the large amount of plants to choose from at the garden center may overwhelm you. There are innumerable high-light plants that are tempting, yet probably will not do well in your home. I want you to find a plant that will work in your situation so that your plant is healthy and you are successful in adding beauty to your home. I want you to look at your plant every day and be happy you chose it because it is thriving in your home. Hopefully, by reading the previous chapters you have determined the amount of light you have to offer a plant so that it will not only survive but thrive. Though many of the plants in this chapter can tolerate low-light situations, some would do best in a medium light. In the previous chapters, we discussed ways to improve the light you have, and even how to add electric lights to expand the range of plants you're able to grow. Remember, with electric lights, you can grow a plant anywhere.

The following plant profiles will help you find and choose a plant that will do best in the conditions you have to offer. There *is* a plant that will work for you. If you have enough light to read a book, there is a plant that will live in your home. Choose one you find attractive. I beg you not to run out and buy a plant just because it is all the rage online. That plant may not be the best one for your situation and you and your plant could both be frustrated and unhappy. So, read on and find a plant that will fit your light situation and that will make you happy.

ANGEL VINE

LOW LIGHT

WET

TOXIC TO PETS

Other Common Names

Mattress vine

Botanical Name

Muehlenbeckia complexa

Angel vine's diminutive leaves and wiry stems give an airy look to any planter and are suitable for a hanging basket. This plant also works well as a groundcover around a specimen plant in a large pot, as long as the container companion needs the same light and water preferences.

Light

Angel vine prefers bright light but can tolerate and grow moderately well in a lower light. If it is placed in a lower-light setting, it won't grow as exuberantly and also won't use as much water, which could be a plus. An east or west exposure is best, but a north window would work well too.

Water

Keep this vine well watered. Because of its thin leaves, it dries out more easily, and all the leaves may fall off. If you are lucky to observe the problem soon after it has dried out, rehydrating it may bring it back—even if all the leaves have fallen off, it may still be alive. Let it sit for a bit and you may be surprised when the vine sprouts new leaves.

Size

The small-leaved vine may seem dainty and diminutive, but it is a rampant groundcover in its native habitat. In our homes it may reach 3 to 4 feet long or even more.

Propagation

Place root-tip cuttings in a moist potting medium. The leaves are thin, so covering the cuttings with plastic or glass will keep the humidity up while they are rooting.

Pet Safety

Toxic to dogs and cats.

'ANITA' DRACAENA

LOW LIGHT   DRY   TOXIC TO PETS

Botanical Name

Dracaena reflexa 'Anita'

Dracaenas as a group are wonderful, easy-to-grow plants. 'Anita' is no exception. With narrower leaves than those of the most common dracaenas, it has an airy, less coarse look. This plant cleans the air of harmful volatile organic chemicals (VOCs). It is often sold as a standard or small tree form.

Light

Like other dracaenas, 'Anita' can tolerate low light levels, such as those provided by a north-facing window, but will thrive in a bit brighter light as offered by an east or west window. It would do well a few feet from a south window as well, but do not place in direct or high light, as it may burn the leaves.

Water

Dracaenas do not like to be overwatered or placed in "heavy" potting medium; they prefer a well-drained, porous medium. Water your dracaena well until water runs out of the drainage hole, then let the medium dry down at least halfway before watering again.

Size

You will find 'Anita' in smaller-size pots that are economical compared to the tree form. The tree form is a good focal point in a room where a lot of light isn't available. It can get quite tall, so if needed, it can be trimmed back to stay shorter and bushier.

Propagation

If you decide to trim your 'Anita', you can insert the cuttings into a container of moist potting medium to start a new plant. Remove the bottom leaves from the cutting, inserting only about 1 inch of bare stem into the medium for rooting purposes.

Pet Safety

Toxic to dogs and cats.

ARROWHEAD VINE

MEDIUM LIGHT

MOIST

TOXIC TO PETS

Other Common Names

Arrowhead plant, goosefoot

Botanical Name

Syngonium podophyllum

Seeing a young arrowhead plant at the store, you might not assume it would become a vine in its mature form. If you would prefer it to stay more compact, you can keep it trimmed back when it starts to vine. This characteristic may take a while to develop. The plant's most attractive factor is its arrowhead-shaped leaves, which may range in color from silver to green to pink and any or all of those colors mixed together. It is an easy plant to grow.

Light

Place in bright light and turn frequently, as it tends to lean toward the light in a short span of time. If placed in a too-bright light, the leaves may sunburn. An east, west, or north exposure is preferable. If all you have is a south exposure, place it far away from the window.

Water

Keep the arrowhead evenly moist and place it on a pebble tray for extra humidity. Allowing the plant to dry out or allowing the humidity to drop too low may cause the leaf edges and tips to brown. Because of its thin leaves, it may dry out more quickly than some of your other plants.

Size

Before vining, the plant will be only 12 to 18 inches tall; when vining, it could reach up to 3 feet. The vines can be trained to grow up a mossy pole or trellis or can be trimmed to stay smaller. There are newer varieties that have been hybridized to stay small, under 6 inches, to be used in fairy or miniature gardens.

Propagation

Take 6- to 8-inch tip cuttings and root in a moist potting medium. Because of the plant's thin leaves and love of humidity, covering the cuttings while rooting may be helpful.

Cultivars

'Moonshine'—This variety is a light silver in color.

'White Butterfly'—One of the most popular varieties available, with light-green leaves edged in dark green.

'Pink Splash'—A variety with pink markings scattered on medium-green leaves.

'Mini Pixie'—A miniature white-and-green variety that may only grow to 3 inches, perfect for a terrarium or fairy garden.

'Pink Fairy'—A miniature pink variety that also stays approximately 3 inches tall.

Pet Safety

Toxic to dogs and cats.

BIRD'S NEST ANTHURIUM

LOW LIGHT   MOIST   TOXIC TO PETS

Other Common Names

Cabbage anthurium

Botanical Name

Anthurium plowmanii

If you need a bold statement in your room, a bird's nest anthurium is your plant. The name may lead you to believe that this is a smaller plant, but it's more on the scale of an eagle's nest than a sparrow's. The large, leathery dark-green leaves arise from the soil with only a short petiole or stem, growing in a circular arrangement—a rosette form, like a nest. The leathery leaves are a good defense against insects, as they are difficult to chew.

Light

Give these plants a medium light, such as an east or west window; they would also do well in the low light of a north window. If enough light is present, they will flower, but it isn't all that attractive. A small rattail-like spadix will arise from the base of the leaves, but it is not showy like its cousin *Anthurium andraeanum*, which we are more used to seeing with the bright colorful spathe surrounding the spadix.

Water

Keep your anthurium evenly moist, not allowing it to dry out completely; use a pebble tray to raise the humidity around it. In addition, keep it warm—these plants do not appreciate temperatures below 50°F.

Size

The leaves can grow over 3 feet long.

Cultivars

'Ruffles'—A form with leaves with ruffled edges.

'Fruffles'—A form with extra ruffles on the edges of its leaves.

Pet Safety

Toxic to dogs and cats.

BIRD'S NEST FERN

LOW LIGHT   MOIST   SAFE FOR PETS

Other Common Names

Nest fern

Botanical Name

Asplenium nidus

If you envision ferns as having long fronds with many little leaflets equally spaced along either side, this fern will surprise you. Its fronds are an entire leaf, with no leaflets present. The fronds form a bowl shape with a brown fuzzy "nest" in the center. The new fronds arise out of that brown nest in an oval shape that resemble small eggs held in the nest, thus the plant's common name. Keep the fronds free from dust by using a sponge to wipe the dust off.

Light

Place this fern in medium light; its favored exposure is an east window, but it will grow in a north window as well. It can also be placed a few feet from a west window or many feet from a south window. Direct sun may burn the fronds.

Water

Never allow any fern to dry out completely, yet do not let it stand in water either. Raise the humidity by setting it on a pebble tray. Water around the edge of the pot, trying not to get too much moisture in the center of the "nest," as it may rot the plant and cause its fronds to fall off. In its native habitat, the bird's nest fern often grows as an epiphyte on trees. As it would be growing on an angle on the tree, water never sits in the middle of the "nest" for too long. In our homes, planted in a container, they are grown straight, so excess water in the middle of the plant may not drain completely.

Size

These ferns get quite large in their native habitats and can grow 4 feet high and 3 feet wide. In your home, they most likely won't get that large.

Pet Safety

Not toxic to pets.

BIRD'S NEST SNAKE PLANT

LOW LIGHT   DRY   TOXIC TO PETS

Other Common Names

Dwarf snake plant, good luck plant

Botanical Name

Sansevieria trifasciata 'Hahnii'

While there are many kinds of snake plants, this type forms short rosettes of leaves resembling the round shape of bird's nests. They come in a range of colors, from dark green to bright yellow, some with stripes and patches of variegation. These small plants are perfect for low- and medium-light areas. The more variegation on the leaves, the more light the plant will need to keep its colors vibrant.

Light

Snake plants are known for their tolerance of low light, especially the dark-green varieties. If given a medium to bright light, however, they will do better and multiply rapidly.

Water

Keep this plant on the dry side, especially in a low-light situation. If kept too wet, it will result in the complete collapse of the plant due to rot. For the same reason, do not leave water sitting in the middle of the rosette. If you keep the plant in medium to bright light, water when the medium is almost completely dry. If the leaves become wrinkly, like your fingers after you've sat in a bathtub too long, it is in dire need of water.

Size

Bird's nest snake plants can range from 4 inches to more than 1 foot high. They will expand with the growth of offsets from the base of the parent plant.

Propagation

The easiest way to propagate this plant is to separate the offsets from the parent and plant them in their own container. A single leaf can also be cut into pieces, allowed to callus over, and be planted in a moist medium. Make sure to place the leaf so that the original bottom of the leaf is in the medium or else it will not grow. If you are trying to propagate a snake plant with a yellow edge, be aware that the yellow edge will not be present on the new plant.

Cultivars

'Golden Hahnii'—A golden variety with stripes of lighter yellow and green. This plant will not exceed 5 inches in height, but after a few years it will send out offsets and may reach 8 to 10 inches around or more if not separated.

'Black Star'—Dark green leaves edged with yellow.

'Jade'—A pure dark-green variety.

'Starlite'—Gray leaves edged with yellow.

Pet Safety

Toxic to dogs and cats.

BLUE STAR FERN

LOW LIGHT   MOIST   SAFE FOR PETS

Other Common Names

Bear's paw fern

Botanical Name

Phlebodium aureum

This fern's blue color makes it a showstopper. As you get closer, your attention is then stolen by the huge "caterpillars" creeping along the potting medium. These furry rhizomes (spreading stems) crawl across the surface of the medium, sending up fronds along the way. Blue star ferns are normally found on trees, growing as epiphytes in their native habitats. The thick, leathery fronds are more forgiving of the low humidity in our homes than most ferns.

Light

This plant needs low to medium light. Set it in an east or north window or back a couple of feet from a west window.

Water

Though it is forgiving of low humidity, it still benefits from the humidity of a pebble tray. Do not let the potting medium dry out—keep it evenly moist.

Size

This fern can get quite large in its native habitat but will probably not exceed 2 feet high and wide in your home. The rhizomes will crawl up and over the pot rim when they hit the edge, so a low, wide pot is best for this footed fern.

Propagation

The spores that appear on the back side of the fronds can be sown in a moist medium and covered to keep the humidity high. A faster and easier way to propagate this fern is to remove a piece of the rhizome with a frond attached and pin with a bent piece of wire to a container of moist potting medium. Keep it moist to allow the new roots to form.

Pet Safety

Not toxic to pets.

BOSTON FERN

LOW LIGHT

MOIST

SAFE FOR PETS

Botanical Name

Nephrolepis exaltata 'Bostoniensis'

You may notice this fern growing outside in the summer, as many people use it as a hanging plant on their porches. I have had a cultivar of this fern for over 30 years, passed down to me from my great-grandmother, and it has exclusively been growing in my home. The biggest complaint about this fern is the messiness of falling leaflets from the fronds. They do drop some leaflets as part of the normal aging process, yet this is a beautiful plant and worth a little clean-up. Place it on a pedestal to show off its airy, arching fronds.

Light

Ferns in general love a medium light. They can tolerate low light, such as in a north window, but the medium light in an east window would be its preference.

Water

Keep this fern evenly moist, never allowing it to dry out. Use a potting medium that has plenty of peat but good drainage. Place on a pebble tray to raise the humidity. It will drop fewer leaflets if well cared for and never allowed to dry out completely; keep the humidity elevated.

Size

Large for a fern, this one can become 3 feet tall and wide.

Propagation

This fern can be divided and potted up individually. It may also be propagated from the long stolons it sends out. Pin them to a container of moist potting medium while still attached to the parent. New plants may already form on the stolon before you pin them to moist media, making it easy to start a new plant.

Cultivars

'Suzy Wong' (cotton candy fern)—A fluffy, foamy fern that, true to its name, resembles swirls of cotton candy. This is a newer variety that also needs plenty of water and humidity.

'Rita's Gold'—This bright chartreuse fern was discovered by Rita Randolph; she gave a piece to Alan Armitage, who trialed it and named it 'Rita's Gold'. This is an exceptional variety that adds a bright spot to any room and is also used extensively in shady combination containers outside.

'Tiger Fern'—A cultivar with striking variegation on the fronds.

Pet Safety

Not toxic to pets.

BRAKE FERN

MEDIUM LIGHT

MOIST

SAFE FOR PETS

Other Common Names

Silver ribbon fern,
table fern

Botanical Name

Pteris cretica

This is another fern you may
walk by without realizing it
is a fern—a unique species
that has fronds with one to
five pairs of pinnae (the primary
division of a fern frond) that
look more like ribbons than
the leaflets normally seen on
ferns. It mixes well with
other ferns, as it has such a
different look.

Light

A medium to bright light is
best, such as an east window or
back a bit from a west window.
A north exposure is fine for
nonvariegated forms.

Water

As with most ferns, keep it evenly
moist, never allowing it to dry
out or stand in water. Raise
the humidity by placing it on a
pebble tray.

Size

This is a smaller fern and grows
1 to 2 feet tall and wide.

Propagation

The plant may be divided
to propagate.

Pet Safety

Not toxic to pets.

BUTTON FERN

LOW LIGHT   MOIST   SAFE FOR PETS

Other Common Names

Round leaf fern

Botanical Name

Pellaea rotundifolia

This is an endearing small fern with little leaflets that resemble small buttons. The leaflets grow from arching fronds with stems of dark brown. This plant looks great in a hanging basket. The dark green, round leaflets are thicker than most ferns and so can better tolerate the low humidity of our homes. It also is fine in lower-light conditions.

Light

Place this fern in low to medium light. An east window is best, but a north window would also work.

Water

This fern wants to be kept evenly moist, but not wet. Water the fern and then let the potting medium dry down a bit before watering again.

Size

From 6 to 12 inches tall and wide.

Propagation

The fern can be divided and potted up individually.

Pet Safety

Not toxic to pets.

CAST-IRON PLANT

LOW LIGHT   MOIST   SAFE FOR PETS

Other Common Names

Bar-room plant

Botanical Name

Aspidistra elatior

This plant was popular in Victorian times, as it survived in dark, drafty, cold parlors. Because of this tolerance of adverse conditions, it was dubbed the cast-iron plant. It is perfect if you have a low-light situation and less than stellar moisture and temperature conditions. There are variegated forms, but they need more light to stay variegated.

Light

The cast-iron plant can tolerate low light but does well in medium light. The newer variegated cultivars need a medium light to keep their variegation. Do not place this plant in direct sun, as it will burn.

Water

It is also known for its tolerance to drying out but would prefer to be evenly moist. The less light it has, the less water it will need.

Size

The long, strappy leaves can be up to 2.5 feet long. Wipe them down with a moist sponge to keep the dust to a minimum, optimizing the light it can collect.

Propagation

Separate sections of the plant and plant them up individually.

Cultivars

'Milky Way'—A speckled variety (shown).

'Variegata'—A white-striped variety.

'Snow Cap'—A variety with white-tipped leaves.

Pet Safety

Not toxic to pets.

'CEBU BLUE' POTHOS

LOW LIGHT

MOIST

TOXIC TO PETS

Botanical Name

Epipremnum pinnatum
'Cebu Blue'

This new cultivar of the ubiquitous pothos family barely resembles its relatives. Its blue leaves are a striking color not usually seen in houseplants. If it receives too much light, it can appear washed out and sickly. This is a vine that definitely does not want to be in too-bright light, and direct sun will burn its foliage. If it is in lower light levels, it will be a deeper blue. If you are looking for something different from the usual pothos, this one is for you. As the plant matures, the leaves will become lobed and larger, though it may not mature in your home setting. It is also called the blue philodendron but is a pothos, not a true philodendron.

Light

Place this plant in an east or north exposure. If the leaves seem overly light colored, move to a spot with less light.

Water

Keep the potting medium evenly moist. This cultivar has much thinner leaves than the usual pothos, so it will not tolerate being allowed to dry out.

Size

This vine will become as long as you allow it to grow, but would look better if the longer vines are trimmed so that the plant stays full. Cut some long tendrils back to the potting medium level and new tendrils will grow.

Propagation

Root tip cuttings in water or in moist potting medium.

Pet Safety

Toxic to dogs and cats.

CHINESE EVERGREEN

LOW LIGHT

DRY

TOXIC TO PETS

Other Common Names

Aglaonema

Botanical Name

Aglaonema spp.

This easy-to-grow plant used to be offered to the public only in a green form splotched with darker-green markings. Now, it is being hybridized faster than one can keep track of, and boasts pinks, reds, and peach colors too.

Light

The older hybrids, mostly green colored, can take low light and grow well. The newer colorful hybrids need medium light and do well in an east or west window. If placed in low light, they will lose their bright coloration.

Water

Let the soil dry down 1 to 2 inches before watering. Chinese evergreens also prefer higher humidity, so place them on pebble trays.

Flowers

Aglaonemas will flower in good light, but these plants are grown for their beautiful foliage; it benefits the plant to cut off the flowers so its energy goes to foliage production. The flowering spadix is surrounded by a white spathe.

Size

Ranges from 12 inches to approximately 3 feet.

Propagation

Aglaonemas can be propagated by stem cuttings or division.

Cultivars

'Anyamanee'—Usually sold as red aglaonema in the stores. It has variegated dark pink leaves and grows 12 to 15 inches tall.

'Creta'—Green leaves with red markings, 12 inches tall (shown).

'Emerald Beauty'—This is one of the older varieties that can take low-light conditions. It has dark-green leaves with light-green mottled stripes and grows up to 24 inches tall.

'Silver Queen'—Another older variety but with the opposite coloring from 'Emerald Beauty': light-green leaves with dark-green mottled stripes. It also can take lower-light conditions and grows up to 18 inches tall.

'Pink Dalmatian'—A beautiful cultivar with pink splashes on shiny dark-green leaves that grows up to 12 to 18 inches tall.

'White Lance'—The leaves of this unusual cultivar are only 1 inch wide and a light gray color. It grows 18 inches tall.

'Sparkling Sarah'—This cultivar sports a pink midrib with pink veins on a bright-green leaf and grows 12 to 15 inches tall.

Pet Safety

Toxic to dogs and cats.

CORN PLANT

LOW LIGHT · DRY · TOXIC TO PETS

Botanical Name

Dracaena fragrans

If you've ever driven past a cornfield or picked corn on the cob, you can clearly see this plant's resemblance to a corn stalk. It is tall and has strappy leaves. The corn plant is quite often used in office settings, as it can tolerate low light and some neglect—yet, if well taken care of, it makes a dramatic statement in any room. Keep the long leaves dusted and clean for a more attractive specimen. When purchasing this plant, you will notice it usually has three different heights of woody stems, supporting a fountain of green at each tip. They are sold like that to make for a fuller container of greenery.

Light

As noted above, this plant can tolerate low light but would prefer medium to bright light, though not full sun—direct sun will burn the leaves. Place in an east, north, or west window, or back quite a few feet from a south window.

Water

It is important to water evenly over the whole potting medium to avoid rotting the corn plant's canes. These canes can have small root systems and may need to be straightened after traveling home as well as later as they settle in. Be careful when straightening the canes not to firm the medium too much, which will compact it and force the oxygen out. As they grow, the root systems will get larger and be better able to support the canes.

Size

This plant can reach 6 feet or more in height.

Propagation

There are three ways the corn plant can be propagated. First, the tip of the plants can be cut and rooted to make new plants; this may become necessary to keep the plant at a shorter height. Second, you can cut the tall brown cane to a shorter size, and new growth should push out of the sides of the cane near the top. Third, allow the piece of cane you cut off to dry a bit and then place it on a moist potting medium and keep warm. Make sure the bottom of the cane is the part touching the medium and roots should form.

Cultivars

'Massangeana'—A yellow stripe runs down the middle of the leaves (shown).

'Victoria'—The leaves of this cultivar are shorter and wider and have bright-yellow stripes. The variegated leaves will need more light to keep their variegation.

Pet Safety

Toxic to dogs and cats.

CREEPING FIG

LOW LIGHT

MOIST

TOXIC TO PETS

Other Common Names

Climbing fig

Botanical Name

Ficus pumila

This creeping plant is well suited to a hanging basket. Its crinkled leaves come in dark green or white and green, and it is often used as a groundcover in warm climates. It's also a common terrarium plant due to its thin leaves that need higher humidity.

Light

Give varieties with dark-green leaves low to medium light. If the plant is variegated, it will need a brighter light to keep its variegation.

Water

Never let this plant dry out; it will drop leaves and may not recover. Do not let it stand in water either, but keep it moist. Because it has thin leaves, it needs high humidity, so place it on a pebble tray.

Size

These vines grow almost flat to the surface of the medium but may spread many feet. Keep them pruned to control the size.

Propagation

Insert stem cuttings in a moist potting medium.

Cultivars

'Quercifolia' (oak-leaf creeping fig)—This small cultivar has leaves shaped like oak leaves, hence the cultivar name (*Quercus* is the genus name for oak trees). Because of its small size, it is quite often used as a fairy-garden groundcover.

'Snowflake' (variegated creeping fig)—This cultivar has a green leaf with white edges.

Pet Safety

Toxic to pets.

CROCODILE FERN

MEDIUM LIGHT   MOIST   SAFE FOR PETS

Other Common Names

Alligator fern

Botanical Name

Microsorum musifolium

When you see this fern, you will understand why it got its reptilian name. Its fronds are long and strappy and have a seersucker-like texture that resembles crocodile skin. It is often found growing as an epiphyte in its native habitat, high in trees.

Light

A medium to bright light is best. This fern does well in an east exposure but would also do well in a north window. Avoid direct sunlight on the fronds, which may burn or bleach them.

Water

Plant this fern in a well-drained, peat-based potting medium. Keep your fern evenly moist but never standing in water. Place it on a pebble tray to keep the humidity high and never allow the plant to completely dry out.

Size

Though this fern can grow 4-foot fronds in its native habitat, in the home it will rarely surpass 2 feet.

Propagation

These plants can be divided and planted up separately.

Pet Safety

Not toxic to pets.

DIEFFENBACHIA

MEDIUM LIGHT

MOIST

TOXIC TO ALL

Botanical Name

Dieffenbachia spp.

The beautiful markings on the leaves of these plants, even in lower-light situations, make them a popular houseplant. Most of the cultivars boast large leaves with splotches and patches of darker green, white, or yellow, sometimes all on the same plant. Keep this plant away from children and pets, as the plant's sap contains calcium oxalate crystals that can burn the mouth and throat and may cause a temporary paralysis of the vocal chords.

Light

Place in medium to bright light, such as an east or west window, or many feet from a south window. The less variegated forms do fine in the low light of a north window.

Water

Keep this plant evenly moist and raise the humidity by placing the container on a pebble tray.

Size

Cultivars range from under 1 foot tall to 4 to 5 feet tall.

Propagation

Cut the top few inches off a stem and root in a moist medium. The stems or canes can be cut into pieces, each with a node, and laid horizontally on a moist medium to root.

Cultivars

'Camille'—A bright-chartreuse leaf with dark-green edges.

'Camouflage'—A bright-chartreuse leaf with splotches of dark green.

'Sterling'—A medium-sized plant that has dark-green leaves with chartreuse midribs and veins running through them.

'Tropic Snow'—This large variety can grow to 5 feet or more in height and has a bright-green leaf with a yellow middle feathering into the green edges.

'Tropic Honey'—This large variety has all-yellow leaves with thick dark-green edges.

Pet Safety

Toxic to pets (and people).

ENGLISH IVY

LOW LIGHT

MOIST

TOXIC TO PETS

Other Common Names

Ivy, European ivy

Botanical Name

Hedera helix

Ivy is very versatile, which accounts for its popularity. It looks beautiful in a hanging basket or as a simple vine flowing down from the top of a bookshelf or refrigerator. It comes in a range of cultivars too, with large or small leaves in many combinations of green, yellow, and white. The variegated types will need more light to maintain their color.

Light

The plain-green varieties of ivy can tolerate low light levels, but the variegated ones prefer medium to bright light.

Water

Plant your ivy in a well-drained potting medium. Water thoroughly and then allow the mix to dry slightly before watering again. Never let it stand in water. An evenly moist medium is best for the ivy. If it gets too dry, the roots may die back and be unable to take up water when it is applied; this can lead to a complete collapse of the plant. Keep the humidity high around your ivy with a pebble tray, as dry air around an ivy is like a dinner bell calling spider mites to come and get it. When you do water ivy, take it to the sink and spray with the sink sprayer, cleaning the ivy leaves and deterring spider mites.

Size

Ivy can grow long stems but can be kept under control by trimming.

Propagation

Pot stem cuttings in moist potting medium. You can also pin the stems to another container of potting medium while staying attached to the parent. Roots will form where the stem touches the moist medium. When the roots are established, the stem can be cut from the parent and grown on its own.

Pet Safety

Toxic to pets.

FAIRY WASHBOARD

MEDIUM LIGHT   DRY   SAFE FOR PETS

Botanical Name

Haworthia limifolia
var. *ubomboensis* (left) and
Haworthia limifolia (right)

If you are obsessed with
succulents but have problems
growing them because you don't
have enough sun, there is hope!
Haworthias are the perfect
succulents for the low-light
conditions of our homes. They
grow in small rosettes and
each leaf has ridges protruding
from the surface, thus giving it
its common name. This small
succulent rarely reaches more
than 4 inches across, making
it perfect for an indoor fairy
garden—especially appropriate
given its name.

Light

Place this succulent in a low to
medium light. Do not give it the
full sun other succulents prefer,
as it will turn burgundy and
may sunburn.

Water

These succulents need the
potting medium to become
almost completely dry before
watering again, especially if they
are growing in lower light.

Flowers

The flower stalk will appear from
the center of the rosette and may
extend over 2 feet with small
white trumpet-shaped flowers.

Size

The small rosette of leaves will
be approximately 2 inches tall
and not more than 4 inches wide.

Propagation

Remove offsets from the base of
the plant and pot up individually.

Cultivars

There are many haworthias
to choose from. All require less
light than the usual succulent,
so look for the *Haworthia*
genus, not necessarily just the
limifolia species.

H. limifolia var. *stricta*—The
ridges on this variety are white.

Pet Safety

Not toxic to pets.

GRAPE IVY

LOW LIGHT

MOIST

SAFE FOR PETS

Other Common Names

Oak-leaf ivy

Botanical Name

Cissus rhombifolia 'Ellen Danica'

You've probably heard the warning "Leaves of three, let it be." While this plant does resemble poison ivy, it fortunately doesn't cause an itchy rash. A cascading plant, grape ivy is perfect for a trellis or hanging basket. It has dark green leaves that are separated into leaflets.

It is robust and can cover a problem spot or a dark corner quickly.

Light

This versatile vine can tolerate a low-light north window but would prefer a medium light in an east exposure or back a few feet from a west window.

Water

Plant in a peat-based but well-drained medium and keep the plant evenly moist, not allowing it to stand in water.

If it dries out too much, the leaves will turn brown and drop.

Size

This vine can reach lengths of 10 to 12 feet. Trim it to keep it a more manageable size if needed.

Propagation

Root tip cuttings in a moist potting medium.

Pet Safety

Not toxic to pets.

GREEN WORM FERN

MEDIUM LIGHT   MOIST   SAFETY UNKNOWN

Other Common Names

ET fern, grub fern

Botanical Name

Polypodium formosanum

Some people think "footed" ferns are a bit creepy. The creeping rhizomes, which are actually modified stems, on this particular version resemble green worms or grubs. The airy light-green fronds arising from the feet, though, make it a beautiful plant and a conversation piece. The rhizomes will creep across the container, crawl over the edges, and keep growing.

Light

A medium light, preferably an east window, is best for this fern as with others, but it could do well in a north window as well. Or place it a few feet back from a west window—the sunlight will be too intense if the plant is too close to the window.

Water

Keep this fern evenly moist. If it dries out, the fronds will begin to dry and fall off. Because of the succulent nature of the rhizomes, if it is watered regularly, new fronds will grow—even if they previously dropped due to drying out. Raise the humidity by setting the container on a pebble tray.

Size

The rhizomes will creep to the edges of the container, climb over the rim, and keep growing. A low, wide container is best for this fern. The fronds will rise above the foliage approximately 12 to 18 inches.

Propagation

Take cuttings of the rhizome with a frond attached. Pin them to a container of moist potting medium with a florist pin or piece of bent wire.

Pet Safety

Unknown.

HEARTLEAF PHILODENDRON

LOW LIGHT

MOIST

TOXIC TO PETS

Other Common Names

Sweetheart vine, heartleaf, parlor ivy

Botanical Name

Philodendron hederaceum

The heartleaf philodendron and the pothos are probably tied as the most popular houseplants of all time; the heart-shaped leaves and ease of care account for their popularity. Today's newer cultivars keep that popularity high. The dark-green leaves allow them not only to survive in a low-light situation but to thrive.

Light

This philodendron does well in the low light of a north window but thrives in medium light, such as an east window or back a few feet from a west window. A southern exposure may burn or bleach the leaves.

Water

This is a forgiving plant if it dries out but would prefer to be kept evenly moist. It does not want to be wet, though.

Size

This trailing plant can get quite long, but you can keep it bushy by trimming some of the stems back to the soil line. New shoots will emerge.

Propagation

Root stem cuttings in a moist potting medium.

Cultivars

'Brazil' (pictured)—A cultivar with dark-green leaves with bright-green accents.

'Lemon Lime'—A cultivar with bright-green leaves.

P. brandtianum—Gray leaves with dark-green veins.

P. micans—Dark-green leaves that appear to be quilted, not flat like the heartleaf philodendron.

Pet Safety

Toxic to pets.

HOLLY FERN

LOW LIGHT   MOIST   SAFE FOR PETS

Other Common Names

Japanese holly fern

Botanical Name

Cyrtomium falcatum

The shiny dark-green leaflets on the fronds of the holly fern do slightly resemble holly leaves, with the pointed notches along their edges, but the plants are not related in any way. Because the fronds are a little bit leathery, this fern is more forgiving than others of the dry air in our homes.

Light

Give this fern low to medium light, such as an east or north window, or set back a few feet from a west or south window.

Water

Keep it evenly moist, not allowing it to dry out nor letting it stand in water. Though it is forgiving of dry air, it would be better to set the fern on a pebble tray to raise the humidity.

Size

The fronds can be up to 2 feet long, so the fern can be 4 feet wide.

Propagation

The plant may be divided to propagate.

Pet Safety

Not toxic to pets.

JAPANESE ARALIA

MEDIUM LIGHT   MOIST   SAFE FOR PETS

Botanical Name

Fatsia japonica

The Japanese aralia is a great focal point in a room. Its large, palmately lobed leaves have seven to nine lobes, and the plant can grow to several feet high. It is often sold as a single stem, which grows tall with its large leaves protruding on beefy stalks. It can be trimmed to make it branch out and become more shrublike.

Light

For best growth, place this plant in medium light, such as an east or west window. It will also do well in a low-light setting, such as a north window, or back a number of feet from a south window.

Water

Keep the *Fatsia* plant evenly moist. If it dries out, it may drop its lower leaves. Keep the humidity up around the plant and keep it away from heat registers, as the plant will attract spider mites if it is kept too dry and warm air is blowing on it.

Size

It can become a stately plant that, in its native habitat in Japan, could be up to 15 feet tall. In the home it will most likely grow only to 6 feet.

Cultivars

'Spider's Web'—A cultivar with white mottled foliage that is attractive but needs more light than its all-green counterpart.

Pet Safety

Not toxic to pets.

JEWEL ORCHID

MEDIUM LIGHT   MOIST   SAFE FOR PETS

Other Common Names

Black jewel orchid

Botanical Name

Ludisia discolor

Though the white spikes of flowers on this plant are beautiful, they are nothing compared to its amazing foliage, whose burgundy color with iridescent peach stripes make this terrestrial orchid an unusual plant. It is an extremely easy orchid to grow in potting medium and blooms in a medium-light window. These plants can become leggy as they get older. Take tip cuttings and, after rooting them, plant them back in the pot to make for a fuller and more attractive plant.

Light

Jewel orchids are found in shady places in their native habitat, so they are perfect for our homes. They do need bright light to bloom, though, so give them medium light such as that from an east window. Turn it regularly to promote flowering on the entire plant. This plant can grow well in a lower light, though it won't have any blooms.

Water

Use a heavy, peat-based potting medium and keep it evenly moist.

Flowers

The small flowers are white and appear on flower stems that rise above the foliage approximately 12 inches.

Size

The foliage is only a few inches high, but the stems do extend over the edge of the pot and hang down approximately 8 to 10 inches. These plants make an excellent hanging basket.

Propagation

Root tip cuttings in moist potting medium. The plant can also be cut apart and pieces potted up individually.

Pet Safety

Not toxic to dogs or cats.

KANGAROO FERN

MEDIUM LIGHT   MOIST   SAFETY UNKNOWN

Other Common Names

Kangaroo paw fern

Botanical Name

Microsorum diversifolium

The shiny bright-green fronds of this fern are deeply lobed. It is a "footed" fern but is different from other footed ferns in that its rhizomes aren't overly fuzzy and are a dark chocolate color. The kangaroo fern is usually offered as a hanging basket; the rhizomes will keep right on growing over the rim and down the side of the container. They could completely cover the pot, if allowed, or they can be moved into a wider pot.

A low, wide pot is the best container for this fern, so it can creep across the potting medium.

Light

As with most ferns, it likes medium light—an east window is perfect. It could also work in a north window, a few feet back from a west window, or even further from a south window.

Water

Do not let this fern dry out; keep it evenly moist, never wet. If it dries out, the result will be yellowing leaves that will eventually fall off, though it is a little forgiving because the rhizomes hold a bit of water.

Keep the humidity up by placing the container on a pebble tray.

Size

The fronds rise approximately 1 foot above the rhizomes, and the plant will spread as wide as its container and beyond.

Propagation

Remove a piece of the rhizome with a frond intact and pin it to a moist potting medium.

Pet Safety

Unknown.

LEMON BUTTON FERN

LOW LIGHT   MOIST   SAFE FOR PETS

Other Common Names

Fishbone fern

Botanical Name

Nephrolepis cordifolia

This close cousin of the Boston fern (see page 69) is much more diminutive than its large relative. It is also called fishbone fern because of the placement of the leaflets on its fronds. Many confuse this fern with the button fern, *Pellaea rotundifolia* (page 73) because of its round leaflets. The leaflets of the lemon button fern are much thinner than the button fern's, so the plant needs higher humidity. When its leaves a re crushed, it gives off a slight lemon scent.

Light

Place your fern in low to medium light; light that is too bright will burn the fronds. This is a perfect small plant for a desk at work if there is enough light to support it. The fluorescent lights in your office would probably be enough to keep it growing well.

Water

Water your fern regularly to keep it moist. Do not allow it to completely dry out or it will lose its leaflets. As with almost all ferns, the lemon button would like high humidity around its fronds, so place it on a pebble tray to raise the humidity level.

Size

Up to 12 inches tall.

Pet Safety

Not toxic to pets.

'LITTLE HOPE' PHILODENDRON

LOW LIGHT

MOIST

TOXIC TO PETS

Botanical Name

Thaumatophyllum bipinnatifidum
'Little Hope' (formerly
Philodendron bipinnatifidum)

Do you love the look of the
huge split-leaf philodendron
(*Thaumatophyllum selloum*,
formerly *Philodendron selloum*) or
the swiss cheese plant (*Monstera
deliciosa*, see page 113) but don't
have the room for such an
imposing plant? The 'Little
Hope' philodendron may be
right for you. It has the split-leaf
philodendron look but is easy to
care for and small enough to fit
in a small apartment or home.

Light

As with many philodendrons,
this plant will tolerate low
light well. Medium light
would be preferable, but it
is a versatile plant and adapts
well to its situation.

Water

Keep the potting medium
evenly moist, letting the
top couple of inches dry
out before watering again.
The amount of light the plant
receives will be important
when deciding how much
water it needs.

Size

It will grow to less than 2 feet
and may be 3 feet around at
maturity. In comparison to the
regular split-leaf, it is considered
a dwarf plant.

Pet Safety

Toxic to pets.

LUCKY BAMBOO

MEDIUM LIGHT   WET   TOXIC TO PETS

Other Common Names

Ribbon plant, curly bamboo, Chinese water bamboo

Botanical Name

Dracaena sanderiana

Lucky bamboo—which is not a bamboo at all, but a dracaena—has been popular since it first came to the market in the late 1990s. It is said to bring luck and is used extensively in feng shui. If the stem is in a curled formation, it has been trained that way by being turned to the light—a form of phototropism. It is also available with braided stems or in other trained forms.

Light

Dracaenas need medium to bright light to grow best. In lower-light situations, they may stretch for the light but will still do well.

Water

This plant is most often grown exclusively in water but can be grown in a potting medium as well. Dracaenas do not like the chemicals in municipal tap water, so if possible use rainwater or distilled water. Change the water at least one to two times per month and keep it at the same level all the time. If you grow this plant in soil, keep it evenly moist.

Size

Canes can range from 1 inch to many feet tall, according to how long the grower has cut them. They can be kept trimmed.

Propagation

Root cuttings in moist potting medium or place them in water to grow roots. If you cut the top of the cane off, new sprouts will emerge lower on the cane. The cane piece that has been cut off with the green leaves still attached can be placed in water to grow new roots.

Pet Safety

Toxic to pets.

MONSTERA

MEDIUM LIGHT

MOIST

TOXIC TO PETS

Other Common Names

Swiss cheese plant, fruit salad plant

Botanical Name

Monstera deliciosa

This midcentury-modern decorating staple is once again one of the most popular houseplants in the world. The high ceilings and open-concept floor plans popular today have brought this large plant back into the spotlight, helped by its large architectural presence and ease of care. The huge perforated, lobed leaves are unique and are thought to have holes in them to combat the strong winds and large amounts of rain they can be exposed to in their native environment, high in the trees of the rainforest. They send out aerial roots to gather more moisture and to stabilize themselves—do not allow them to attach themselves to your wood floors or other surfaces, as they will leave marks when pulled away.

Light

Monstera can tolerate low light but prefers medium to bright light. In its native habitat, it starts life on the jungle floor and scrambles along until it finds a tree to cling to, then climbs to the top for light.

Water

Keep this plant evenly moist, letting it get quite dry before watering again.

Size

This plant can get large and will need a lot of room to grow. Growing it on a moss pole is best so it has something to cling to, and it may reach 10 or more feet tall.

Propagation

Plant stem-tip cuttings in moist potting medium, or root them in water. The plant can also be air-layered.

Cultivars

'Variegata'—A variegated form that has splotches of light green and white on the dark-green leaves.

Pet Safety

Toxic to pets.

MOSES IN THE CRADLE

MEDIUM LIGHT

DRY

TOXIC TO PETS

Other Common Names

Oyster plant

Botanical Name

Rhoeo spathacea

The two bracts that surround this plant's small white flower resemble shells, thus the common name of oyster plant—and of course the white flower cradled down in the bracts brings to mind Moses floating down the river in his basket. This plant, used for a groundcover in the south, makes a good low- to medium-light houseplant. The upward-pointing leaves allow the purple undersides to show. The tops of the leaves are dark green.

Light

Nonvariegated cultivars of this plant can take low to medium light, such as a north or east window. If your plant is variegated, medium to bright light is recommended, such as an east or west window.

Water

Keep this plant evenly moist. Do not overwater it, as it easily rots, and it is better to err on the side of too dry rather than too wet. It appreciates a higher humidity to ensure the tips do not turn brown.

Flowers

A small white flower surrounded by two bracts deep in the leaves of the plant, apparent only when you think to look for them.

Size

Grows 1 to 1½ feet tall.

Propagation

Remove offsets and pot them up separately.

Cultivars

'Tricolor'—This newer, popular cultivar is bright green with white and pink stripes, and the undersides of the leaves are bright pink.

'Vittata'—An older cultivar that has yellow stripes on the green tops of the leaves but still boasts the purple undersides.

Pet Safety

Toxic to pets.

NERVE PLANT

MEDIUM LIGHT   MOIST   SAFE FOR PETS

Other Common Names

Mosaic plant, silver net plant

Botanical Name

Fittonia spp.

The nerve plant's beautifully veined leaves are its main attraction. The smaller-leaved varieties of this plant have thin leaves that will require a higher humidity than our homes usually offer; a terrarium is a perfect environment for them. The plants come in pink, white, green, and red, and some even have piecrust edges. The smaller forms of this endearing plant are also used often as a fairy garden plant.

Light

Low to medium light is best. High light will burn the leaves.

Water

This plant does not want to be too wet, as it will rot. On the other hand, do not allow it to dry out, as it will drop its leaves. Keep it evenly moist. This plant loves high humidity, so place it on a pebble tray or in a terrarium.

Size

'Pink Wave' can reach 10 to 12 inches tall, but the smaller versions will be low-growing groundcover up to 4 to 5 inches tall.

Propagation

Root tip cuttings in moist potting medium.

Cultivars

'Pink Wave' (pictured)— A large-leaved variety grown for its leaves that are thicker than the smaller varieties, thus not needing such high humidity to grow well in our homes.

'White Anne'—Green leaves with bright-white veins.

'Red Anne'—Green leaves with red veins.

'Pink Star'—Green leaves with pink veins with ruffled leaves.

Pet Safety

Not toxic to pets.

OX-TONGUE

MEDIUM LIGHT

DRY

SAFE FOR PETS

Botanical Name

Gasteraloe 'Little Warty'

This plant's common name comes from its leaves, which some say resemble a tongue, with their rounded tips and the tubercles (small round protuberances). Ox-tongue is easy to grow and makes a great medium-light succulent houseplant; it can also be placed in a lower light and still do well. This particular variety is a dark green with white tubercles all over the leaves.

Light

Gasterias can thrive on a west or east windowsill. They could also be placed a few feet from a south window and do well.

Water

These fleshy succulents should be planted in a fast-draining medium and never allowed to sit in water. Keep drier in the winter when the light levels are lower.

Flowers

The name *Gasteria* comes from the plant's flowers, which resemble the shape of a stomach. Ox-tongue will easily produce these flowers on a west windowsill. Most are orange with a green tip and hang from 2- to 3-foot-long stems. This gasteraloe is a cross between the genus *Gasteria* and the genus *Aloe*, and its flowers are more tubular, like the *Aloe* plant.

Size

Gasterias can range from a little over 1 inch high to more than 2 feet tall, depending on the variety.

Propagation

These plants make quite a large number of offsets, which can be removed and potted up individually. They also can be started from seed. Single leaves can be removed, allowed to dry for a few weeks, and planted in a moist potting mix. Or they can be laid horizontally on the medium and they will grow babies and roots from the cut end in a few months.

Pet Safety

Not toxic to pets.

PARALLEL PEPEROMIA

MEDIUM LIGHT   MOIST   SAFE FOR PETS

Other Common Names

Radiator plant

Botanical Name

Peperomia puteolata

The parallel peperomia, though not a technically a vine, does become a weeping plant with age and may be found for sale in a hanging basket. The leaves are whorled around the red stem in groupings of three to five and are heavily veined, with the veins running parallel to the edges of the leaves, thus the common name. This family of plants is quite easy to grow as long as you do not overwater, as most are quite succulent—whether it be their leaves or stems that are succulent.

Light

Give this plant a medium-light setting, such as an east or west window. It will do well in a lower-light situation but may stretch for the light.

Water

Plant in a well-drained potting medium and keep it evenly moist, but not wet.

Size

Parallel peperomia can grow up to 15 inches long.

Propagation

Root tip cuttings in a moist potting mix.

Pet Safety

Not toxic to pets.

PARLOR PALM

LOW LIGHT   MOIST   SAFE FOR PETS

Botanical Name

Chamaedorea elegans 'Bella'

In Victorian times, this palm graced nearly every parlor, thus earning its name. Like the cast-iron plant (page 75), it could survive the dark, cold conditions of the homes of this era. Its tolerance to low light has been the main reason for its popularity. It grows slowly, and tiny versions of this plant are often used in dish and basket gardens.

Light

The parlor palm can tolerate low light levels but would prefer a medium light. If it receives too much light, its normal bright-green color will turn yellowish.

Water

Keep the soil evenly moist, but do not let the plant stand in water. Make sure to use a good, well-draining potting medium. To lessen the chance of a spider-mite infestation, keep the humidity high by placing the plant on a pebble tray and giving it a refreshing shower of water at least once a month—this also keeps the dust off the leaves.

Size

This plant slowly grows to 3 to 4 feet high.

Propagation

Propagate from seed.

Pet Safety

Not toxic to pets.

PEACE LILY

MEDIUM LIGHT

MOIST

TOXIC TO PETS

Other Common Names

White sail plant, spath

Botanical Name

Spathiphyllum spp.

Related to the aglaonema, philodendron, and dieffenbachia rather than to true lilies, these popular plants are easy to care for and forgiving of underwatering. The white flowers appear with medium light, which the plant prefers. The beautiful, shiny, dark-green leaves are very attractive.

Light

The peace lily can tolerate medium to low light, but flowers may not appear in a low light. An east or north window would work well, with the east window producing flowers. It will flower 5 to 6 feet away from a west window.

Water

This plant does not like to dry out, so keep it evenly moist. Although it will wilt from underwatering, it comes back quite well as soon as water is added to the medium. Some use the wilting as a visual indicator to water, but if that happens often, leaves will start dying back from the tips and yellow leaves will appear. It is better to check the plant often and keep it moist.

Flowers

While it looks like a flower, the large, white flag-like appendage of the peace lily is actually a spathe. The white upright cylinder appearing in the middle of the spathe is the spadix and is covered with tiny flowers. Pollen falls from them, dusting the leaves with a powdery white substance. In commercial settings, the spadix is quite often removed to keep the leaves clean and pollen free. The spathe and spadix last for a long time, which is a nice bonus. As they age, they will turn brown and at that time can be removed by cutting the stem as close to the potting medium as possible.

Size

Peace lilies have many cultivars, ranging from 1 foot tall to over 4 feet.

Propagation

The peace lily is a multiple-crown plant. The easy way to propagate these is to just separate the crowns and plant them up individually.

Cultivars

'Domino'—A variegated form with white markings on the puckered leaves.

Pet Safety

Toxic to pets.

PEACOCK MOSS

MEDIUM LIGHT   MOIST   SAFE FOR PETS

Other Common Names

Spike moss, club moss, frosty fern

Botanical Name

Selaginella kraussiana 'Variegata'

While neither ferns nor true mosses, selaginellas are considered "fern allies" and like similar growing conditions. The iridescence of the leaves of this low-growing plant makes it attractive to collectors. The white-tipped peacock moss variety is often sold at holiday time and called the "frosty fern." There are also red-colored varieties for added interest. Because of its moisture and humidity preferences, the frosty fern makes a great terrarium groundcover.

Light

Medium light is best, as in an east window. These plants do not want to be in high light, as it will bleach them out. The lower light of a north window would work as well, but the white-tipped variety needs more light to remain variegated.

Water

Keep this plant evenly moist, never allowing it to dry out. If it is grown as a holiday plant, it may come in a foil cover. Remove the cover when watering so that the plant is never standing in water. Raise the humidity by placing it on a pebble tray or by growing in a terrarium. These plants never want to completely dry out, so check them often. If the top of the potting medium feels dry, give it a drink of water.

Size

This is a small plant, only a few inches high, though being a groundcover it can spread quite far.

Propagation

Propagate by spores or division.

Pet Safety

Not toxic to pets.

PEACOCK PLANT

MEDIUM LIGHT   MOIST   SAFE FOR PETS

Other Common Names

Zebra plant, rattlesnake plant

Botanical Name

Calathea makoyana

The beautiful markings on this plant's leaves give it its colorful common names—the light-green leaves have dark green stripes and spots on top, and the undersides are a burgundy color. The spotted pattern reminds us of peacocks, and the stripes of zebras. Calatheas need even moisture and high humidity.

Light

Place in medium light, such as an east window or back a few feet from a west window. This plant will also do well in a north window.

Water

Keep the humidity high by placing these plants on pebble trays and, if possible, growing them in a bathroom or kitchen window, where the humidity is a little higher already. Keep the well-drained potting medium moist, but not wet. Never let calatheas dry out completely. Dry plants, dry air, and fluoride in tap water can all cause brown edges and tips on the leaves.

Size

Up to 2 feet tall or more.

Propagation

Propagate by division.

Pet Safety

Not toxic to pets.

PHILODENDRON MAYOI 'TAHITI'

LOW LIGHT

MOIST

TOXIC TO PETS

Botanical Name

Philodendron mayoi 'Tahiti'

Like most philodendrons, this plant is very versatile to light conditions. 'Tahiti' will most likely be found for sale in a hanging basket. It does well in that form but would prefer to grow up a trellis or a moss pole, which you can find at garden centers or online. The moss pole is inserted into the plant's container and the tendrils cling to it; this mimics its natural habitat of growing upwards on trees in the jungle.

Even a foraged dead branch can work, and may be more interesting—or you can make your own moss pole with chicken wire and moss. When this plant grows upward, the size of its leaves usually increases. If you grow it on a support, especially a branch, it can be a great conversation piece in the room.

Light

Place your philodendron in medium to low light. Do not place it in direct sun, as it may burn or bleach the foliage.

Water

Keep it evenly moist, but not too wet or standing in water.

Size

This plant will only be approximately 12 inches tall in a hanging basket, but the vines will take off and be as long as you let them grow. Keep it trimmed back to keep the size under control.

Pet Safety

Toxic to pets.

PIGGYBACK PLANT

LOW LIGHT

MOIST

SAFE FOR PETS

Other Common Names

Mother of thousands

Botanical Name

Tolmiea menziesii

The endearing thing about this unusual plant is the way little plantlets grow on top of its leaves. Each small plantlet is a replica of the mother plant. This plant is native to the northwestern part of the United States, where it grows as a shade groundcover in damp woods and along creeks. It spreads from underground stems.

Light

This plant can tolerate a low-light setting but would like to be in medium light. Do not place it in direct sun, as it may burn.

Water

Do not let your piggyback plant dry out completely. It may recover, but the leaf edges will be brown and crispy.

Size

This plant will form a mounded shape up to 12 inches tall. In its native habitat, it can grow up to 3 feet, but in the house it most likely will not become that large.

Propagation

Remove a leaf with a leaflet growing on it plus a small piece of stem and place it in a container of moist potting medium.

A plastic cover is helpful to keep the humidity up until it grows roots. Or leave the leaf attached to the mother plant and pin it with a hairpin or bent piece of wire, such as a paper clip snipped in half, to a small pot of moist potting medium. It will root, using the food and nutrients coming from the mother plant. After it roots, you can cut it away from the mother.

Pet Safety

Not toxic to pets.

PIN STRIPE CALATHEA

MEDIUM LIGHT · MOIST · SAFE FOR PETS

Other Common Names

Pin stripe prayer plant

Botanical Name

Calathea ornata

The beautiful foliage of this plant is its main attraction—who wouldn't love its pink pin stripes? It is often considered to be a prayer plant, and it is in the same family as the *Maranta* genus (page 143). The pink stripes feather out from the midrib to the edges on a slight curve over the dark-green leaves. The undersides of the leaves are burgundy.

Light

Calatheas need a medium exposure to ensure the stripes stay a bright pink. Direct sun will fade the markings, yet too little light will not allow them to stay bright and colorful.

Water

Keep this plant evenly moist—not wet, but never dry either. Place it on a pebble tray to keep the humidity high. This is a must, as the leaf edges will turn brown if grown in dry air. They are also affected by the fluoride in water if your community adds it. Use rainwater or distilled water to ensure the edges and tips aren't brown.

Size

This calathea may grow to 2 feet tall.

Propagation

Propagate by division.

Pet Safety

Not toxic to pets.

PLUM PINE

MEDIUM LIGHT   MOIST   TOXIC TO PETS

Other Common Names

Buddhist pine, southern yew

Botanical Name

Podocarpus macrophyllus var. *maki*

In much of the southern United States, *Podocarpus macrophyllus* is grown outside as an evergreen hedge, much like the yew of northern climates—thus the common name of southern yew. The variety *maki* is most often used as a houseplant, as it stays more compact with shorter leaves. It can be kept even smaller with pruning or made into a shaped topiary or even a bonsai. If left to grow naturally without pruning, it takes on a weeping appearance.

Light

This plant prefers a medium to bright light but can tolerate low light as well.

Water

Keep it evenly moist and plant in a fast-draining potting medium. Do not allow it to sit in water, as it is susceptible to root rot.

Size

The plant grows up to 6 to 8 feet but with pruning can be kept 4 to 5 feet tall. It is an excellent large floor plant.

Propagation

Take tip cuttings, dip them in rooting hormone, then plant them in moist potting medium. This plant can also be grown from seed.

Pet Safety

Toxic to pets.

POLKA DOT PLANT

MEDIUM LIGHT   MOIST   SAFE FOR PETS

Other Common Names

Freckle face plant

Botanical Name

Hypoestes phyllostachya

Polka dot plants have gained in popularity in the last few years because they are easy to find and easy to grow, and they have bright, colorful foliage. Often found in the fairy gardening section at your local garden center, these plants are also popular in dish gardens and terrariums. They are thin-leaved and need high humidity levels to grow their best. Place them in a window over the kitchen sink or in the bathroom, use a pebble tray, or place them in a terrarium for best humidity.

Light

Medium light is best, so place polka dot plants in an east window or back a few feet from a west window. If they don't have enough light, they may stretch. If placed in an area where they get full sun exposure, the thin leaves may become washed out or even burnt.

Water

Keep this plant evenly moist, not allowing it to dry out, as it quite unforgiving of dry soil. Low humidity will cause brown leaf tips and edges.

Size

It is important to keep these small plants trimmed so that they stay full. They can reach 16 inches or more and become quite leggy and sparse if not trimmed.

Propagation

Root tip cuttings in moist potting mix or grow from seed.

Pet Safety

Not toxic to pets.

POTHOS

LOW LIGHT

MOIST

TOXIC TO PETS

Other Common Names

Devil's ivy, golden pothos, money plant

Botanical Name

Epipremnum aureum

If there is any plant out there that everyone is familiar with, it has to be the ubiquitous pothos. You can find it framing windows, sprawling down furniture, or spanning beams on the ceiling. Its tolerance to low-light situations is a plus in the houseplant world. These plants are often used in office settings, because they can thrive where the only light they receive is from fluorescent lights.

Light

The golden pothos, which has green leaves with yellow marbling, can tolerate the low light of a north window but would prefer the medium light found within a few feet of a west or east window. If your plant loses its yellow color and reverts to all green, move it into more light and it will regain its variegation.

Water

This plant will let you know it is dry by wilting over the edge of the pot, but it would be best if that never happens, as it will develop some yellow leaves. Keep it evenly moist, but never standing in water, as the roots may rot and the plant will collapse.

Size

In its native habitat, this vine can climb 40 to 70 feet up a tree. In our homes, the vines can become 10 to 20 feet long if left untrimmed, but there may only be leaves on the ends of the stems and the rest left naked. It is better to keep your plant trimmed and full. Cut a few of the stems back to the soil line and new sprouts will appear.

Propagation

Root stem cuttings in water or in potting medium.

Cultivars

'Marble Queen'—This cultivar has exceptionally attractive white- and green-splotched leaves. It will need more light than some varieties, as it has a lot of white on the leaves, but not full sun, which will burn those white parts.

'N' Joy' and 'Pearls and Jade'— These cultivars are similar, with more organized white- and-green patches. 'Pearls and Jade' also has small dots of green softening the edges of the different colors. Both are newer exceptional hybrids.

'Neon'—A bright-chartreuse green cultivar that brightens any room with its color.

Pet Safety

Toxic to pets.

PRAYER PLANT

MEDIUM LIGHT   MOIST   SAFE FOR PETS

Other Common Names

Rabbit tracks, herringbone plant

Botanical Name

Maranta spp.

Imagine having a plant that folds its leaves up at night, quietly rustling them in the process. This plant gets its common name from that interesting habit, as it appears to fold its "hands" in prayer. The main attraction, though, is the beautiful look of the leaves. Some have red stripes and splotches, while others have dark-green splotches on a lighter-green leaf. The undersides of the leaves on some species are burgundy, adding to the interest of this striking plant.

Light

Medium light, such as an east window, is preferred. You can also place this plant back a foot or so from a west window, a few feet from a south window, or close to a north window.

Water

Keep the soil evenly moist and avoid fluoridated water—used bottled or distilled instead. Do not allow these plants to dry out. They can be a little bit finicky, as their leaves may develop brown edges if the humidity is too low, so use a pebble tray to keep the humidity high.

Size

Prayer plants stay relatively low, usually under a foot, but can spread out 2 to 3 feet if grown well. As larger plants, they are often offered for purchase in a hanging basket.

Propagation

Tip cuttings should root in a moist potting medium.

A large plant may also be separated into smaller plants.

Cultivars

M. leuconeura var. *erythroneura*—The leaves of this variety have light-green splotches running the length of the midrib and bright pinkish-red veins curving away from the center. A beautiful version of the prayer plant.

M. leuconeura var. *kerchoveana* (rabbit tracks) (shown)—The leaves on this variety have dark-green splotches on either side of the main midrib. The variegated version (*M. leuconeura* var. *kerchoveana variegata*) has the same dark-green splotches but also white and red splotches that make it more interesting.

Pet Safety

Not toxic to pets.

'SILVER SATIN' POTHOS

LOW LIGHT   DRY   TOXIC TO PETS

Botanical Name

Scindapsus pictus 'Silver Satin'

I cannot say enough about this easy, low-light plant. I have one growing in my bathroom and one in my living room, both set back far from the windows, and they are doing great in the lower light. The thick leaves mean the plant doesn't need to be watered often, especially in a low-light situation. The silver splotches on its medium-green leaves make it an attractive plant. It can be grown with long tendrils to train around a window. You can also place it in a macramé planter or let it cascade off a shelf.

Light

Place this plant in low to medium light and watch it thrive.

Water

Because of its thick, leathery leaves, it stores water well and does not need it as often as thinner-leaved plants. Let it dry down between waterings, but not until it is bone dry. Do not leave it standing in water either.

Size

This plant can produce long tendrils, but it can become more stringy and bare the longer the tendrils become. To keep the plant fuller, cut a few of the tendrils back to the soil line and it will sprout new stems, keeping the middle of the plant full and more attractive. Use the cuttings to make new plants.

Propagation

Root cuttings in water or in a container of moist potting medium.

Cultivars

'Jade'—This variety can take even lower light than 'Silver Satin' because of its solid dark green color. Its thick leaves make it quite drought tolerant as well, though it should never completely dry out.

Pet Safety

Toxic to pets.

SNAKE PLANT

LOW LIGHT

DRY

TOXIC TO PETS

Other Common Names

Bowstring hemp plant, mother-in-law's tongue

Botanical Name

Sansevieria trifasciata

The snake plant is back in vogue. Previously, they were disregarded because they have been commonly placed in dark corners and left to languish, leaves falling over due to lack of proper light. Now, they're back, touting many varieties and forms, plus a reputation as an air purifier. These qualities, along with better information about growing this plant, have given it the place in the houseplant world it deserves.

Light

The snake plant (especially the darker green varieties) will tolerate low-light situations and do well if not overwatered there.

However, they prefer medium to bright light.

Water

Many a snake plant has been killed by overwatering. Water infrequently in a low-light setting, letting the potting medium become quite dry before watering again. In a high-light setting, they will use more water. Never allow a sansevieria to sit in water.

Size

These plants can range from a few inches to many feet tall, depending on the cultivar.

Propagation

Because of the multiple crowns this plant produces, separating is the simplest way to propagate them. Leaves cut into 2- to 3-inch sections and planted upright will form new plants at the base. Make sure the

sections are planted with the top side up or they will not grow.

Cultivars

S. cylindrica—Instead of flat leaves, this snake plant's leaves are round and very sharply pointed. It can grow to over 6 feet tall.

'Bantel's Sensation'—The bright white-and-green striped leaves of this cultivar are very striking. It grows 3 to 4 feet tall.

'Laurentii'—This green-striped variety with yellow edges up the sides of the leaves is a very popular, old cultivar. It grows 3 to 4 feet tall.

S. masoniana 'Mason's Congo'—The leaves of this large variety can be 8 to 10 inches wide, and the plant can grow 3 to 4 feet tall.

Pet Safety

Toxic to pets.

SPIDER PLANT

LOW LIGHT

MOIST

SAFE FOR PETS

Other Common Names

Airplane plant

Botanical Name

Chlorophytum comosum

The charming spider plant is one of the most popular houseplants. The variegated version is usually the one offered for sale and is often in a hanging basket. The miniature plants floating in the air attached to long stems from the parent are the most endearing characteristic of this plant. The tuberous root system means it will need to be up-potted or divided when the roots fill the pot, or it may break the container.

Light

The solid-green version of this plant can live in low light, but the variegated versions need medium to bright light.

Water

Keep the spider plant evenly moist. Brown tips appear with the salt buildup from fertilizing the plant; flush the plant often to rectify the problem and trim the leaves to remove the brown tips.

Size

These plants are 1 to 2 feet tall, but the stems cascade 2 to 3 feet over the edge of the container.

Propagation

The little plantlets at the ends of the stems can be removed and rooted in a moist potting medium or in water. To ensure faster rooting, leave the babies attached to the parent plant and pin them to a container of moist medium. When they are well rooted, cut them away from the parent. Divide a large plant into smaller pieces and pot individually.

Cultivars

'Bonnie'—This cultivar has curly leaves.

Pet Safety

Not toxic to pets.

VELVET CALATHEA

MEDIUM LIGHT   MOIST   SAFE FOR PETS

Other Common Names

Fuzzy pheasant feather

Botanical Name

Calathea rufibarba

The look of this plant's leaves alone make it worth seeking out—they are dark green and lance-shaped with burgundy undersides and wavy edges. If you run your hands up the backs of the leaves, you will find out what makes it even more unique: they feel like velvet. You can't help but pet the leaves like you would your favorite cat or dog. The species name *rufibarba* is derived from the Latin *rufus*, meaning "red," and *barba*, meaning "beard."

Light

This plant does great in a medium light, especially an east window. It also likes a north exposure. Do not place it in a light that is too bright, as it will burn or bleach the leaves.

Water

Keep it well watered, not letting it dry out. It also will need high humidity to deter spider mites from moving in. Place it on a pebble tray to raise the humidity.

Size

This plant will grow to approximately 12 to 20 inches tall.

Propagation

This plant has multiple crowns and can be propagated by division.

Pet Safety

Not toxic to pets.

WANDERING JEW

MEDIUM LIGHT   MOIST   TOXIC TO PETS

Other Common Names

Spiderwort

Botanical Name

Tradescantia zebrina

This vine is a popular hanging basket plant. The attractive striped leaves shimmer in the light with iridescence. The plant is easy to grow, and because of its succulent stems it can take a bit of neglect. The stems are brittle and break off easily, but this is a perfect opportunity to start some more plants and share with others.

Light

This plant does best in medium light; it tolerates low light but may lose some of its color. An east or west window is best. Placing it in a too-bright situation will bleach or burn it.

Water

Keep the potting medium evenly moist, but not too wet, or stem and root rot can set in. Because the stems are succulent, it is forgiving of drying out temporarily.

Size

This vining plant typically doesn't grow taller than 6 inches but can trail over the container by 2 feet or more.

Propagation

Root stem cuttings in a moist potting medium.

Pet Safety

Toxic to pets.

WATERMELON VINE

LOW LIGHT   MOIST   SAFE FOR PETS

Botanical Name

Pellionia pulchra

The watermelon vine's leaf veins resemble a watermelon rind—and to add to the effect, it also has reddish stems. This plant is usually sold as either a small plant for a terrarium or as a large plant in a hanging basket. It is easy to grow, but though it's called a vine, it is more of a trailing plant.

Light

This unique plant needs only low to medium light to be happy. Do not place it in bright light, as it may bleach or burn the leaves.

Water

Watermelon vine appreciates an evenly moist potting medium. If it becomes too dry, it will lose its older leaves. It does not want to stand in water either.

Size

The plant is only a couple of inches high and spreads or hangs down from the container about 1 foot. If it gets leggy, pinch the ends off and propagate new plants to fill in the pot.

Propagation

Root tip cuttings in a moist potting medium.

Pet Safety

Not toxic to pets.

ZZ PLANT

LOW LIGHT

DRY

TOXIC TO PETS

Other Common Names

Zanzibar gem

Botanical Name

Zamioculcas zamiifolia

If you have a dark corner in which every other plant has perished, this is the solution for you. The ZZ plant has become one of the most popular houseplants in the past few years, as it can take low light and still look amazing. It has shiny dark-green leaves with a strong architectural appearance. The leaves are upright and are made up of many leaflets on each rachis (the stem of a compound leaf), the actual "stem" being the underground tuberous rhizomes.

Light

While it is true this plant can tolerate low light levels for quite some time, it prefers medium to bright light to grow to its full potential.

Water

This plant has been touted as extremely drought tolerant, and while it can take long periods between waterings, the frequency depends on the light level it is growing in. If it dries down too much, it will drop leaflets.

Size

This plant can reach up to 3 feet tall.

Propagation

The unusual part of this plant is that it can grow new plants from an individual leaflet, but it takes quite a long time. Place the cut end into moist potting medium and cover with plastic or glass. This process may take many months. The plant can also be divided.

Pet Safety

Toxic to pets.

Index

About the Author

Lisa is the Houseplant Guru, who features all things houseplants on her blog **www.thehouseplantguru.com**. She grew up in rural mid-Michigan, where being immersed in nature every day nurtured her love for all things outdoors, especially plants. Living down the road from her grandma meant spending a lot of time there and watching her lavish attention on her African violets and other houseplants. This is where the love for them began.

Being an avid outdoor gardener as well has led to a column for *Michigan Gardening* magazine and frequent articles for *Michigan Gardener* magazine. In addition, she has written for HGTVgardens.com, *Real Simple* magazine, and the houseplant section of Allan Armitage's Greatest Perennials and Annuals app. She lectures extensively around the country, spreading the word about the importance of houseplants and how to care for them, and has been interviewed online, in print, and on TV and radio about houseplants.

Lisa worked for more than a decade at Steinkopf Nursery and Garden Center as the annuals and houseplants manager. She is a member of numerous plant groups, including the Michigan Cactus and Succulent Society, the Town and Country African Violet Society, the Southeast Michigan Bromeliad Society, and the Hardy Plant Society.

She cares for hundreds of houseplants in her home in the Detroit area, where she lives with her husband, John, and two adorable kitties. She loves to visit conservatories during her travels and is a volunteer at the Belle Isle Anna Scripps Whitcomb Conservatory in Detroit. Lisa feels that every home, office, and apartment should have a houseplant, and that there is a houseplant for every situation. A green thumb is something anyone can have because everyone needs a little green in their lives!

Acknowledgments

First, I must thank my editor Alyssa Bluhm for bringing her idea to me and asking me to write this book. She felt there needed to be a book for people who live in low-light situations, helping them find plants that work and learn how to make them happy. I loved working with you! Thank you to the team at Quarto for bringing our ideas to life.

Secondly, I need to thank Heather Saunders for adding her beautiful pictures and unique perspective to the book. It was a joy to work with you! We would both like to thank her sons Harrison and Julian Saunders for their youthful exuberance and might helping move plant friends (and pots, photography equipment, potting medium, plant stands, and more) in and out of locations.

To Danielle Dirks for the usage of her gorgeous Detroit Airbnb units as the featured locations, and also for her assistance with being a model for a few pictures. To Tim Travis and Jim Slezinski for allowing us to use their gorgeous houseplants, containers, and garden center for photo shoots. To Jay Atwater and Chelsea Steinkopf for allowing their plant friends to be supermodels. To Kelly Ardito for her extraordinary photo assistance on this project. To Chelsea Steinkopf for her assistance with photo shoots.

I also would like to thank some wonderful people who were of much assistance with the plant list and the botanical lingo: Justin Hancock from Costa Farms; Brett Weiss, horticulturist from the Vivarium; and Jeremy Kemp from the Belle Isle Anna Scripps Whitcomb Conservatory. To my good friend Nancy Szerlag, who assisted with technical potting medium and fertilizer questions and also with editing.

Of course, thank you to my family and friends for putting up with me while writing another book and all that it entails—forgotten appointments, missed events, and added stress. To my heavenly Father for bringing this opportunity my way and getting me through every minute of every day.

And though he is last, he is the most important: my husband, John, my rock, my biggest fan, and the love of my life. Thank you for putting up with the house that hasn't looked its best since the first book and even more houseplants than before taking over the window space. Did you want to see into the backyard anyway? Love you!